THE
TOTAL
BANANA

ALEX ABELLA

THE
TOTAL
BANANA

HARCOURT
BRACE
JOVANOVICH
·
NEW YORK
AND
·
LONDON
·
AN ORIGINAL
HARVEST / HBJ
BOOK

Requests for permission to make copies of any part of the
work should be mailed to:
Permissions, Harcourt Brace Jovanovich, Inc.
757 Third Avenue, New York, N.Y. 10017

The Banana Alphabet, or "Watch That Peel," by Marc
Miyashiro, 1978.

Printed in the United States of America

Library of Congress Cataloging in Publication Data

Abella, Alex.
    The total banana.

    (An Original Harvest/HBJ book)
    1. Banana.   2. Cookery (Bananas)   I. Title.
TX558.B3A23   1979   641.6'4'722   79-1855
ISBN 0-15-690475-6

First Original Harvest/HBJ edition

A B C D E F G H I J

To Paola:

"Tanto gentile e tanto onesta pare
la donna mia quand'ella altrui saluta. . . ."

CONTENTS

FOREWORD

Why the banana? was a question friends invariably asked when I mentioned this book. Just as invariably I responded, Why not? The banana, after all, is the most popular fruit in America —close to 5 billion pounds were consumed in 1977. Besides, the banana bears considerable weight in our culture; it is used extensively in song, paintings, literature, and all other arts, in addition to being incorporated into the national lexicon as a choice symbol for madness. Furthermore, the attributes of the banana are the very features that distinguish America.

The American knack for mass production, which aims to supply the greatest good to the greatest number, is the earmark of the banana—raised, picked, and distributed according to the latest techniques that provide, year round, a dependable and low-priced food. Moreover, the standardization of taste, appearance, and size of the banana represents our leveling democracy, which has raised homogeneity to the heavenly sphere of the virtues. And, finally, the banana is a fruit of the people: it stands for novelty songs instead of symphonies, pop collages instead of classical oils.

Usually after this argument my listeners, overcome by the barrage, nodded thoughtfully and kept quiet. Of course, I could have told more: about the shiny hands of tiny bananas my father used to bring home, which I gulped with such delight when growing up in Cuba; about my shock at first hearing the term *banana republic*, which brought to mind a land without cities or inhabitants, awash in a tide of yellow fruit; also, about my sudden realization (while bunched in a crowded streetcar) of the split personality of the banana, an object of joy and scorn, the most liked yet most deprecated fruit in America.

When I started researching this work three years ago, I went into it with the assumption that little had been published about the banana. Like most preconceived ideas, that one was false too. There is an immense wealth of literature about all aspects of the fruit, particularly in mass-circulation periodicals, which in recent years have shown a predilection for the banana. The encyclopedic work of N. W. Simmonds, *Bananas*, has been of great assistance. Earlier books by Fawcett, Reynolds, and Kervegant also have proved most helpful in tracing the story of the banana through the ages.

Finally, I wish to extend my gratitude to those persons in the trade who took time from their schedules to answer the meandering questions of a banana neophyte. I am much obliged as well to San Francisco lexicographer Peter Tamony for his suggestions on the association of bananas and insanity. Above all, my thanks to Anna Banana, whose voluminous collection of banana material and flashing insights into the comedy of the banana proved invaluable in the writing of this book.

A. A.
San Francisco
June 1978

THE
TOTAL
BANANA

A Slice
of
Banana
History

HE American love affair with the banana began at a birthday party. The occasion was the 1876 Philadelphia Exposition, a combination World's Fair and centennial ball attended by hundreds of thousands of American citizens and a handful of foreign royalty. Some of the products that would soon help spread the American eagle's wings over the world were previewed—the telephone, the typewriter, the refrigerated car. And so was the banana.

At a booth a sadly propped banana plant displayed tattered leaves to entranced spectators while hawksters sold the fruit, gaily attired in colorful tinfoil, at fifty cents a finger. Some buyers later recalled that the fruit was well past normal ripeness—in fact, rotten. It didn't matter. The banana, with its exotic tropical air, fascinated more for its image than for its reality. Less than twenty years later, $1\frac{1}{2}$ billion pounds a year were being imported to satisfy American banana lovers.

Then, as now, bananas consumed in America were grown in the countries known to traders and archaeologists as Middle America. In those republics the fruit was so plentiful that many experts thought the banana, like the tomato and the cocoa bean, was a native American contribution to the world diet. Even the famed naturalist Alexander von Humboldt affirmed that pre-Columbian Indians knew the pleasures of the banana. But those theories were based on a slip by a Spanish colonial chronicler who had mistakenly placed the banana among his list of Inca foods. Actually, the banana was born in Asia thousands of years ago, migrating with the Spanish conquistadors to America in 1516.

Modern theories argue that the tropical jungles of Malaysia were the starting point of the original banana, which was tough and seedy. The development of the current variety is attributed to the wisdom (or perpetual hunger) of primitive man, who nurtured the seedless type after it arose spontaneously by a lucky stroke of evolution. According to experts, once man stopped roaming and settled down to hearth and family, he turned to fishing and cultivation of vegetative propagation foods: those that can be raised from parts of the plant instead of seeds. The banana fit this Neolithic bill of fare and was probably among the first plants to be domesticated.

The early origin of the banana, predating the invention of the written word, is confirmed in linguistic theory by the proliferation of unrelated terms for the fruit in Asian tongues. For instance, in New Guinea alone over fifty native dialects are spoken, and in each the banana has a different name.

Countless ancient legends were woven around the banana. For instance, in Burma, man allegedly first heard of the fruit from a captured bird that snitched on the banana as a fitter dinner for its jailer. The Hindus were the first to immortalize the banana. Under the name *tala* or *pala*, it is scattered throughout the sacred books of ancient India. Recurring references to the banana pop up in the epics of the Pali Buddhist Canon of 600 B.C., as well as in the *Ramayana* (see the chapter "The Art of the Banana"). Today the banana is still holy, a valued offering to all the gods in the Hindu heaven, particularly to Lord Siva, creator of the universe, and to Kali, goddess of fertility and death. The natives of certain provinces of India also use the banana in festivals and ceremonies; entrances to marriage altars are decked with wreaths of banana leaves, and every spring peasants give bunches of bananas to landowners as a sign of fealty.

Today it is generally believed that Westerners did not stumble upon the banana until 327 B.C., when the conquering armies of Alexander the Great reached the Indus Valley. An account of the encounter was penned by Greek army propaganda chief Megasthenes, who wrote of plants "which in the language of the Indians are called *tala* and which bear fruit in a bunch like those that grow on the crown of date trees." This tale was picked up and embellished by Pliny the Elder. "There is a tree in India," he wrote in his *Historia Naturalis*, "remarkable for the size and sweetness of the fruit, upon which the sages of India live. . . . It puts forth its fruit from the bark . . . a single one containing sufficient to feed four persons. This tree is called *pala* and its fruit, *ariena*."

However, there are other stories about how the banana spread to the West from India and how far it went. Some bananologists conjecture that before Alexander's time traders in skiffs took the fruit to the great centers of civilization in the Middle East. One German historian even postulated the presence of the banana in the Promised Land and in the Old Testament.

The passage mentioned is Numbers 12:23, which takes up the story of the Jews after two years of wandering away from Egypt in the desert, when the Lord relented and sent His people to Canaan. On the way there, an advance party of spies "came unto the brook of Eshcol, and cut down from thence a branch with one cluster of grapes and they bore it between two upon a staff." The historian reasoned that no grapes could ever grow to such dimensions and that the writer of Numbers was obviously speaking of bananas. However, biblical scholars remark that there is no indication in the etymology of the word to suggest that the fruits were anything but grapes, unlikely as that vine may be.

But while experienced hands disagree on the spread of the banana in ancient times, they universally accept the seminal influence of the Arabs during the early Christian era. By the second century Arab traders had opened commercial offices along the coast of East Africa for their flourishing slave and ivory cartel, extending from the Arabian Peninsula to the archipelagos of the South China Sea. During their voyages, these merchants transplanted a number of fruits, among them the banana; native African tribesmen grew fond of the new arrival and carried it inland quickly as they fled from the slave brokers.

The Arabs also carried the banana home to Arabia, not as arid in early medieval times as it is today. One spot renowned for its bananas was Mocha, the city that made coffee famous. The name *Mocha* is now believed to have degenerated into the word *musa*, the Arab designation for the banana and the first name of its Latin alias, *Musa sapientum*. (The second part of that moniker was tagged on in the eighteenth century by the Swedish naturalist and compulsive classifier Carolus Linnaeus, who took his inspiration from Pliny's piece.)

The banana held a prominent position in the Arabic pantheon as the fruit that caused humanity's fall from grace. In the Koran the banana plant is the Tree of Paradise, the equivalent of the Christian Tree of Knowledge, and the banana is the fruit that God forbade Adam and Eve to sample. Turbaned scholars in mosques taught that Eve disobeyed the divine commandment and ate the fruit because the banana so resembled Adam's virile organ that she couldn't resist temptation. Tradition further relates that when the original duo realized their sin, they covered their nakedness with the ample leaves of the banana. A heavenly express then deposited Adam and Eve in Ceylon; from there they traveled to Mecca, taking with them the root of all their troubles, the banana. Part of this legend was picked up by medieval Europe; it is preserved to this day by the French and Italian appellations for the banana, *figue d'Adam* and *fico d'Adamo*.

In the seventh century Muslim armies swept over Africa, the Middle East, and Spain, leaving behind them an easily traced banana trail. A historian who described the rich gardens of Caliph Abd-er-Rahman III in tenth-century Córdoba related that he saw "the plants of Africa mingle with the leaves of the European plants: the palm, the pistachio, the banana growing and developing alongside . . . the olive and the orange tree."

With the further Arab outposts in Morocco and Sudan, the banana traveled farther inland to the heart of Africa and over the next few centuries wended its way to the west coast of the continent. By the time the Portuguese explorers of the fifteenth century reached the coast of Guinea during their search for a trans-African canal, the banana was there waiting for them.

The Portuguese were puzzled by the strange shape and flavor of the banana: the cool climate and gradual Arab retreat from the Iberian Peninsula had destroyed the few plants that had grown roots in Europe. Only a few scholars remembered the purported role of the banana in Paradise; also, time and careless tellers had transformed the banana leaves that covered our first parents' privates into diminutive fig leaves.

The natives of the African coast called the fruit a number of names—*bana*, *gbana*, *abana*, *funana*. They all sounded like *banana* to the tin ear of the Portuguese, who then transported the banana to the Canary Islands, which decades later fell into the hands of Spain. Then, in the year 1516, a Spanish friar, Tomás de Berlanga, sailed from the Canaries to catholicize the heathen Indians of America. Among his Bibles, missals, and mended habits, he stowed the first banana root of the New World. The rhizome, when planted in the island of Hispaniola, grew and multiplied from Mexico down to Peru, Paraguay, and points between. The good father, for his pains and farsighted devotion, was later appointed bishop of Panama.

Upon its arrival in America, the banana completed its circumnavigation of the globe; five centuries earlier, the Polynesians had completed the eastern leg of the journey when they migrated from the Asian mainland to the islands of the South Pacific. From the nineteenth century onward, the history of the banana becomes the story of American business. Yankee traders found it, grew it, and sold it all over the world, transforming the tropical hybrid with a two-week life span into the most common fruit on earth.

The
Medical
Banana

*LEASURE* and goodness often seem the bitterest of enemies, especially when we sit down at the table. Our consciences nag us that our taste for sweets is too pronounced, our weakness for fats, absurd. You may put your conscience to rest when you eat a banana, however. The most popular fruit in America is also one of the most complete foods known to man, woman, and child. It is the fruit for all ages—and many medical conditions as well.

The banana is a low-calorie, low-fat, high-carbohydrate food with many therapeutic and nutritive values. It is used in weight-reducing, low-sodium, low-fat, low-cholesterol, ulcer-preventive, and geriatric diets. Furthermore, the banana is both laxative and paregoric, easily digestible, and supplies vitamins and minerals required for the smooth functioning of the organism.

The average 6-inch-long banana has 100 grams of pulp (about 3.5 ounces) with a remarkably low calorie count—85 calories per fruit, or "finger." In a ripe banana 75 percent of the pulp is water; it is held in a particular molecular structure by fiber and pectin, giving the fruit a fatty texture similar to that of the avocado. Nevertheless, the fat content of the banana is practically nil—0.2 of a gram. A little over 20 percent of the banana is sugar, and the remainder is apportioned among starch, crude fiber, protein, and ash, the last yielding alkaline mineral residues beneficial to digestion.

Surprisingly, bananas constitute an excellent source of vitamin C, providing about one-fourth of the daily amount recommended by the U.S. Department of Agriculture. Bananas also contain a substantial amount of vitamin B_6, useful in preventing perceptual distortions; so you could say that eating bananas keeps you from going the way of the fruit.

Bananas possess an adequate quantity of calcium, niacin, iron, and phosphorus, employed for building red blood cells and aiding proper bone growth. Moreover, they afford an extraordinary supply of potassium, a mineral often drained from the body with the use of diuretic and hypertension-relieving drugs. The body carries only a small amount of this mineral, and deficiencies bring on muscular fatigue, slow reflexes, and clouded mental acuity. In addition, bananas also have minute amounts of manganese, magnesium, zinc, and other minerals. All of these ingredients are encased in a sterile package that resists bacterial infection as long as the peel is intact.

Perhaps the most appealing role of the banana is its performance in a weight-reducing program. Gifted with great bulk, low calorie content, and thick texture for high satiety, the banana is a perfect stand-in for the common snack. For instance, the banana poses a lesser threat to bulging waistlines than one serving of cottage cheese (120 calories).

The weight-reducing properties of the banana were first popularized in the 1930s, when Dr. George Harrop created the "milk and banana" crash diet, a favorite of alcohol-bloated starlets and sedentary entrepreneurs. The regime consisted of six bananas and 1,000 cc (four glasses) of skim milk a day, endowing a mere 960 calories on the body. The combination of the two foods was deemed a nutritionist's dream of perfection, with the milk providing the protein and the banana, the carbohydrates. Harrop modestly recommended his diet on the basis of its "simplicity, low cost, ready availability . . . and demonstrated effectiveness in securing the desired aim."

One of the nutritional myths propounded about the low-calorie banana revolves around its consumption by athletes to gain weight. Undoubtedly, bananas were frequently eaten for this purpose by skinny sportsmen and by patriots during recruiting time for the world wars. Eight to ten bananas were taken between meals, either alone or with milk. However, the bananas were always intended as a *supplement* and not as a substitute for regular meals. If you consider that the bananas of the time—the famous Gros Michel or Big Mike variety—were about 10 inches long and contained 120 calories, it's easy to see how the pounds crept on.

Another nutritional myth of this kind was posed with great feeling by my Italian mother-in-law. "They are indigestible," she said. "I have never eaten one that agreed with me." With filial reluctance, I explained that the supposed indigestibility of the banana arises from consuming the fruit while green and full of starch. The raw banana, I added, is most easily assimilated when totally ripe—that is, when the skin has the brown flecks called sugar spots. My mother-in-law answered, winning still another argument, "But then they are too sweet!"

Of course, sweetness is the greatest attraction of the banana. About 21 percent of the content is sugar, divided into 12.7 percent sucrose, 4.8 percent dextrose, and 3.7 percent levulose. These carbohydrates render the ripe banana sweeter than apples or oranges, which hold only 14 percent and 12 percent sugar respectively. The honeyed flavor of the banana is the work of enzymes during the ripening process, which convert the starch into sugar and a small measure of pectin, the gelatinous substance that makes fruit jelly possible. Besides imparting palatability, these two easily digestible elements also qualify the ripe banana as an apt choice for feeding to infants, a fact known for years by concerned mothers and baby-food companies.

Medical studies have shown that in addition to promoting growth and well-being, the banana is excellent for arresting gastrointestinal disorders in children. The mild laxative properties of the fruit relieve colics and simple diarrhea in infants. These benefits are the result of the fiber and the pectin, which absorb water and produce soft stools while sponging up deleterious bacteria from the intestinal flora. Investigators surmise that the banana also stops these illnesses because it augments the *Escherichia coli* bacteria that aid digestion.

The banana has an unparalleled role in the treatment of celiac disease, a wasting childhood malady that causes a violent rejection of most foods; it is one of the few foods the afflicted child can retain. Although this therapeutic use was accidentally discovered in the 1920s, for decades afterward doctors were at odds about its cause. Only recently was it discovered that infants suffering from celiac disease are acutely allergic to the gluten in many carbohydrates such as wheat and rye. Banana sugars, because of their easy accessibility to the body, are perfect for lessening and ultimately controlling the effects of the disease.

But the banana isn't valuable only to newborns. The stimulation of mineral retention triggered by the banana makes it a sound addition to a young child's diet. Moreover, the banana is a pleasing substitute for the sweets that cause painful cavities and dental bills. For teenagers the banana is incomparable as a slowly metabolized yet steady source of energy, and for adults it can help clear up one of the nation's greatest pains: constipation. The same banana pectin that eliminates a child's diarrhea also unclogs the grown-up's alimentary canal, while the soft fiber and vitamin content promote normal intestinal conditions. This quality is significant to expectant mothers, who often suffer from constipation. Finally, for the elderly the pulpiness of the ripe banana—which can be ingested without mastication—makes the fruit a welcome adjunct to the diet.

Even the soothing effect of the banana, arising from its lack of irritating agents, is a valuable asset; the banana is routinely added to the diet of persons afflicted with colitis, gastritis, and gastroenteritis and is also valuable in the treatment of peptic ulcers because of its capacity for neutralizing hydrochloric acids through alkaline residues (similar to the action of aluminum hydroxide, the main component of Di-Gel, Maalox, and other potent antacids). The banana is often recommended for ulcer patients also because it possesses a high vitamin C content without the accompanying acidic property of citrus fruits such as lemons and oranges.

Yet for more than a million persons in this country the banana is more than a food but actually a guarantee of survival. These are the diabetics who follow a regime of protamine-zinc treatment. Because that remedy has a delayed absorption rate in the body, the diabetic must have a controlled release of carbohydrates into the bloodstream to prevent shock from a low blood-sugar level. Therefore they are regularly advised to eat a bowl of bananas with cream after taking the medicine to prevent the sudden highs and lows in glucose.

But the virtues of the banana are those of omission as well as commission. Its low sodium content makes it a welcome part of the diet of patients who must limit sodium intake because of congestive heart failure, kidney diseases, liver cirrhosis, or other ailments. In fact, the banana is cited as part of the extreme low sodium diet recommended by the Council on Foods and Nutrition of the American Medical Association.

Likewise, its low fat content (with no cholesterol) sanctions the inclusion of the banana in the diet of persons with cardiovascular disease, even of those patients suffering from acute cardiac crises such as myocardial infarctions. Finally, the fruit's low protein level makes it suitable for persons suffering from acute nephritis, uremia, gout, or other kidney diseases.

Some of the therapeutic uses of the banana have been known in foreign cultures for centuries, although in most cases benefits were derived not from the fruit but from the other parts of the plant.

In the eighth century the Chinese physician Men Hsin advocated consumption of the ground banana root or rhizome as a cure for jaundice; other prescriptions declared the effectiveness of the rhizome for curing abscesses, fever, headaches, and measles, as well as for counteracting poisons and aiding in childbirth recovery. Other Chinese pharmacologists touted instead the advantages of banana oil. That liquid, extracted from the bark of the banana plant with a bamboo tube, allegedly cured convulsions of babies, epileptic fits, and stopped "women's hair from falling." No mention is made of its effects on male pates.

In India, wrote a doctor during the 1930s, the juice of the rhizome, because of its great tannin content, was used as mucilage to stop hemorrhages of the genitals. The ashes of the plant were used as an antacid because of their high potassium residue. In nineteenth-century Brazil the sap was considered an antivenom agent for snake bites. It was also said to be handy in the treatment of gonorrhea, leukorrhea, uterine hemorrhage, and inflammation of the larynx.

In the Belgian Congo (today Zaire) European soldiers during colonial times used the sap of the plant to heal the wounds made by the arrows of the natives. One Friar Walker, stationed in Africa during the late 1800s, observed that the Fan tribe of Gabon used "the heart of the banana tree, chopped and steeped in cold water, as a remedy to relieve colics. . . . Sometimes, after having warmed it up over the fire, it is wrapped around the feet of infants to soften the skin and ease the extraction of ticks."

But perhaps the greatest claim ever put forward for the banana was that of the Spanish scientist J. R. Barrera. Writing in 1928, he declared that doctors in São Paulo, Brazil, and in Mexico achieved a 70 percent recovery rate for tuberculosis with a regime of the plant's juice, drunk five times a day, "which achieved many cures in desperate cases." Unfortunately, that remedy could not be duplicated outside Barrera's imagination.

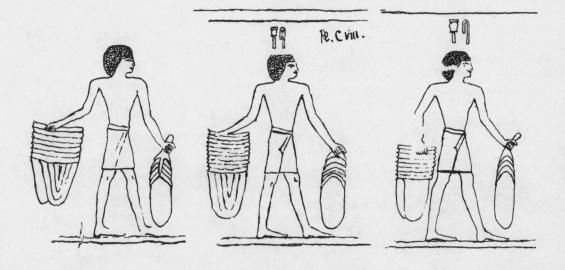

Priests carrying bunches of bananas, from the bas-reliefs of the Temple of Der-el-Bahri in Thebes, Egypt, c. 1500 B.C.

The
Art
of the
Banana

"The most delicious thing in the world is a banana."
Benjamin Disraeli, 1831

HE art of the banana is firmly rooted in cultural history. From the first growths of civilization to the latest blossom of the avant-garde, artists have seized on the banana as an eternally seasonable subject.

The earliest artistic representations of the banana are thought to stem from the Middle East, where bas-reliefs from the temple of Der-El-Bahri (circa 1500 B.C.) at Thebes, Egypt, prefigure the zany spirit later centuries would associate with the fruit. These reliefs show a bunch of bananas being presented as an offering to Hathor, the cow-headed goddess of love, mirth, and joy, by a procession of slyly smiling priests. Savants indicate even earlier possible depictions at Nineveh, the Assyrian metropolis, founded 5,000 years ago, where schematic portrayals of the banana are etched in sculptures and monuments illustrating noble feasts.

But we must look farther east for the first concrete presence of the banana in art—to India, where the banana first slides into view in the frescoes of the temples of Ajanta, painted around 443 B.C. In that country the banana had already been celebrated a hundred years earlier by the poet Valmiki, author of the *Ramayana*, one of the holy books of Hindu civilization.

The *Ramayana*, which sings in countless verses the life and perils of King Rama, lauds the banana as one of the foods that kept the king alive during a long banishment from his realm. Rama's hermitage in the forest, Chitrakusdaram, is also poetized as surrounded by groves of bananas. At the end of the story Valmiki discloses that Rama is one of the incarnations of the god Krishna, and so the banana is deemed holy. Today an ancient song still sung by peasants from the state of Kerala categorically affirms that farmers who don't raise bananas are fools.

In the other great center of non-Western culture, China, the banana was unknown until the second century A.D. because the Chinese kept their civilization in the cool north of Asia. The firsthand Chinese account was set down by one Yang Fu, a Chinese Marco Polo, who called the banana *pa-chiao* and made some quick drawings as a treat for the emperor.

When the Arabs extended their empire in Africa, they thought of the banana. While scientists such as Avicenna and Serapion devoted chapters of their works to the medicinal properties of the banana, poets extolled its sweetness, reminding listeners that existence requires pleasure with practicality. Al-Masudi, an Egyptian bard who died in the tenth century, glorified a delicacy called *kataif*, concocted with almonds, honey, and bananas, widely appreciated by both Muslims and infidels at Cairo and Damascus.

Western eyes were fixed on the banana from the sixteenth century on, after the introduction of the fruit to the Americas. Travelers of the time took pleasure in recounting the odd qualities of the banana. A rhyme on one of these was penned by Thomas Peyton:

Of pleasant taste and sweet delightful hue
If with a knife the fruit in two you reave
A perfect cross you shall therein perceive.

Peyton's sight, or his devotion, was keener than ours; those spots do not make a holy rood.

The transformation of the banana from simple fruit to cultural symbol, with its shedding of the botanical virginity preserved through millennia, took place in the eighteenth century. The culprit was one of the fathers of Romanticism, the French aristocrat and *philosophe* Bernardin de Saint-Pierre, who milked the tears of Europe with his tale of two children growing up in a tropical island, *Paul et Virginie*. In his book *Harmony of Nature*, Bernardin de Saint-Pierre declared that the banana was evidence of the providence of the universe. Why else, he asked, would the banana have a creamy consistency if not expressly for the enjoyment of people of all ages?

Bernardin de Saint-Pierre's treatise confirmed the exotic nature of the banana, permanently coupling it with an image of the tropics, where everything is in perpetual slow motion, where no one needs or desires work, where there is always food for the hungry.

In the United States the banana began to appear regularly in the arts during the nineteenth century. The first sign of our budding affection was given by New Orleans–born Louis Moreau Gottschalk, who in 1844 composed "Le Bananier, Une Chanson Nègre." A great lover of things and people African, Gottschalk wrote "Le Bananier" at age fifteen while studying music in Paris. He derived the rhythms for the song from memories of slave chants. Engaging, with a repeated light motif and many glissandos, the composition sounded extraordinarily exotic and precious; it soon became a trial performance piece for entry to the Paris Conservatory. Gottschalk, the precursor of nationalistic compositions that later gave fame to Dvořák and Glinka, was thus the forerunner of the banana madness that swept the nation around the turn of the century.

After the Spanish-American War of 1898 and the popularization of the fruit, Americans saw the comedy of the banana. Vaudevillians were taken by the banana's phallic shape; other less risqué comedians used it in jokes such as "Remember the banana; when it leaves the bunch it gets skinned" and "The only time some fellows have a girl smile at them is when they slip on a banana peel." Movies featuring slapstick comedians, like those of Mack Sennett and his famous Keystone Cops, brought the lubricant properties of banana skins to new heights; scenes with pompous fools sliding on peels brought down the house. That slickness was so renowned that passengers brought several successful lawsuits against railroad companies for bodily injuries occasioned by slipping and falling on banana skins. An injury of this kind is recorded on the tombstone of Anna Hopwell at Enosburg, Vermont:

> Here lies the body of our Anna
> Done to death by a banana
> It wasn't the fruit
> That laid her low
> But the skin of the thing
> That made her go.

In the more contemplative arts, the banana was handled for its perennial symbolism of torrid climes. Rudyard Kipling mentioned it in his *Jungle Books*, as did E. R. Burroughs in his Tarzan romances. O. Henry in *Cabbages and Kings* and journalist Richard Harding Davis in his novel *Captain Macklin* described the life of the banana-dominated countries of Central America.

The themes of low comedy and politics (which cynics might say are identical) were permanently wedded to the banana during the 1920s and 1930s. But for a spell, at least, comedy triumphed.

No work exploited the comic side of the banana better than the novelty classic "Yes, We Have No Bananas (Today)."

Written by Frank Silver and Irving Cohn in 1923, when Americans seemed to be flouting conventions and avoiding our national weakness for pragmatism, the song won the hearts and ears of the Western world. The tune started such a banana craze that Frank Silver for several years toured the land with a "Banana Band." The few surviving pictures expose the musicians dressed in gold costumes and standing behind daises adorned with glittery banana cutouts.

The song became a fittingly senseless emblem of the age around the world. F. Scott Fitzgerald, the flapper fabulist, in his story "How to Live on Practically Nothing a Year," told how a married couple of American émigrés to southern France were entranced to hear "a song dealing with the non-possession of a specific yellow fruit in a certain otherwise well-stocked store."

In England over half a million copies of the song were snapped up during its first month of sale. Its overwhelming success prompted the suicide of a music-hall composer who drowned himself in the Thames, wailing that he no longer understood the tastes of the British public. The song also made a killing in Germany under the tongue-pleasing title "Ja, Wir Haben Keine Bananen Heute." But Adolf Hitler and the Nazi censors banned it, accusing it of being anti-Aryan propaganda.

The tune had many sequels, the most popular being "I've Never Seen a Straight Banana." In London the song-sheet publishers offered £1,000 to anyone who could produce such a perpendicular fruit. The office was besieged by hundreds of banana-clutching applicants, but since no straight banana was found, the prize was awarded to the least curly exemplar.

Another offshoot of "Yes, We Have No Bananas (Today)" was the rise to fame of black danseuse Josephine Baker, who reaped her reputation in 1924 when she danced to *le jazz hot* at the Folies Bergère with a string of bananas around her waist as sole costume. Baker, whose skin was dubbed "banana colored," became the toast of Europe and was crowned "the personification of expressionism" by excited journalists. Throughout her career Baker resorted to the banana in times of distress; in her autobiography she reflected that her fifty years of stage life had been based on the banana (needless to say, she was a light dancer).

The irrational nature of the banana was next uncovered by the surrealists, who employed it as a weapon in their struggle against academic art. André Breton, who signed the Surrealist Manifesto in 1924, praised the fruit in typically ambiguous terms—"A green banana is as good as a yellow banana," words that in the mouth of anyone else would have made perfectly good sense. Hugo Ball in *Dada Fragments* also got into the act when, hailing the destruction of Western cultural standards, he announced, "Since no art, politics or religious faith seems adequate . . . there remain only the bleeding pose and the banana."

The Crash of '29 put an end to most of that banana nonsense.

In America social unrest and dissatisfaction with business enterprise brought on a wave of muckraking not seen since Teddy Roosevelt bullied his way to power. The previous decade had highlighted the zany side of the banana, but during the 1930s the political association of the fruit gained the upper hand. Bananas and banana companies were routinely denounced as symbols of American despotism in tracts like "The Banana Empire: A Case Study in Economic Imperialism" and even by such mouthpieces of capitalism as *Fortune* magazine.

But the most relentless prosecutors of the banana companies were the Latin American artists who had long screamed against the economic oppression of banana companies. Two Latin American Nobel Prize winners, Miguel Angel Asturias and Pablo Neruda, inveighed against the stranglehold the United Fruit Company had on the life and politics of Central America. Neruda, in his *Canto General*, depicted the fruit powers as satanic entities that caused the spiritual corruption and material putrefaction of the area.

In painting, the representation of the banana was also linked to the fruit companies. Diego Rivera, in his mural *Portrait of America*, drew a panel entitled *Imperialism*. There, to the side of the United Fruit warehouse, a tide of bananas rises to the docks. On top of the crest float baskets of bananas, pineapples, and other tropical fruits; beyond it, banana plants and fields of sugarcane. The scene is framed by the hanged bodies of rebel peons.

By the late 1930s, however, the waggish spirit of the banana was reborn in the figure of Carmen Miranda. Called the Brazilian Bombshell, Miranda was a stock actress in movies featuring the Latin American background popular in that period. Her usual stage demeanor was a doltish wit expressed through flitting eyes and a mischievous grin; her favorite costume was a sheath-like garment enhancing her well-rounded figure, and a fruitbowl hat stuffed with bananas and other tropical fruits. She reached her banana apotheosis in Busby Berkeley's film *The Gang's All Here*, in which a line of chorines paid homage to her by waving giant papier-mâché bananas in her direction.

In 1944, in an apparent imitation of the Miranda bit, United Fruit conceived a publicity campaign around a jingle sung by a flirty banana called Chiquita. The tune became a hit and leaped to the top of the hit parade in a version recorded by Patty Clayton. A group of lonely, admiring soldiers at the war front declared Chiquita "the girl they would most like to be in a foxhole with."

The Second World War saw the first instance of paper currency with a banana design. The bills, circulated by Japanese invading forces at Singapore, were also called banana dollars because of their low purchase value. Most prints were burned by the British after the war ended and the Asian city-state was reclaimed by George VI's empire.

After the CIA-directed coup against the liberal regime of Jacobo Arbenz Guzmán in Guatemala in 1954, Diego Rivera again drew on the banana to represent his political discontent. The most memorable appearance of the banana during the Eisenhower years, however, was in the Harry Belafonte tune

"Banana Boat Song," written by the then unknown Alan Arkin.

Rock music had made a tentative foray at the banana during the 1950s with "Banana, What a Crazy Fruit," but it was up to the age of psychedelia to truly revel in the droll produce. In 1967 British singer Donovan released the album *Mellow Yellow;* a portion of the title tune lyrics foretold a sudden craze for electrical bananas. But instead of the predicted electrical fruit, roasted peels became the rage.

Nineteen sixty-seven was also the year of the hippie "Summer of Love," of widespread experimentation with sex, politics, and drugs. An article published in an underground newspaper divulged that the hidden message of the song was that banana peels gave highs comparable to marijuana highs. "Mellow Yellow" became the amateur doper's call to arms; soon thousands of rebellious youngsters across America were roasting the peels, scraping the insides, and stuffing their pipes with them to inhale the sweet smoke of burning compost. One far-seeing dealer in San Francisco sold only the roasted peel insides, discarding the fruit core as useless.

Sales of bananas jumped, and allegedly spontaneous demonstrations called "be-ins" in New York and San Francisco were held to celebrate the highs of bananas. But the fad expired at year's end after saddened assayers found out that the high was so much banana oil, or bunk. Scientists did confirm that bananas contain serotonin and norepinephrine, substances that modify sensory perception, but that mere combustion of the skins would not produce the altered consciousness desired. The final word was put out by a spokesman from United Fruit, who gleefully asserted that the only high available from a banana is from slipping on the peel.

That same year pop artist Andy Warhol designed an album cover for his rock-and-roll group, the Velvet Underground. Warhol featured, on a white background, a plain yellow banana sticker adhering to the cover• When peeled, the sticker disclosed a pulp painted in the pink tint of human organs. And, in a perhaps unconscious case of banana emulation, the following year comedian Woody Allen wrote, directed, and acted in the first film dedicated to the fruit—*Bananas*—about the adventures of a *schlepp* during a revolution in a tropical country.

Today artists are still finding new angles to the banana. Lukman Glasgow in Los Angeles embraces the fruit because of its satyric overtones and aerodynamic qualities, portrayed in his posters and ceramics. New York painter Larry Blizard metamorphoses the fruit into humans involved in surreally racy situations; and Colombian painter Miguel Botero, San Francisco ceramist G. W. Granizo, and cartoonist Marc Miyashiro use the banana to lend tropical tones to their works.

The art of the banana still grows, by leaps and bounds, in the hearts and minds of America.

The Botanical Banana

HE simplest way to describe a banana, botanically speaking, is by explaining what it is not. In this, as in most instances, common knowledge is myth embroidered by fantasy.

To begin, there is no such thing as a banana tree. The fruit grows from a hapaxanthous herb, a seedy plant without woody tissue that has a single flowering period. But, you say, the banana has no seeds. True, nowadays the banana is also parthenocarpic, or seedless, but that's a recent development that we'll deal with later. For now, remember that this herb or grassy growth is the largest on earth, often soaring to heights of 40 feet from base to crown.

To continue our list of contradictions, the banana plant does not have a trunk but a pseudo stem, composed of tightly wrapped leaf sheaths from which new leaves sprout. Its true stem is what is commonly called the root, but it's not a root; that is, the banana is like a potato, growing from one underground core with many eyes, or bits.

Furthermore, because the banana has no woody tissue, its fruit does not grow on branches. When bearing time arrives, the plant shoots out one single stalk to hold the bunch. When fruiting begins, the banana does not follow gravity's pull but points tenaciously upward to the sky—in effect, the banana grows upside down. And to conclude, the banana is no simple fruit but a berry. That denomination is due to the banana's fleshy pulp, or pericarp, which is supposed to surround the seed, as in currants, grapes, and tomatoes.

But the banana has no seeds! Well, yes. And no. The varieties of banana on the market do not bear seeds because they are hybrids, much like mules, that, although useful, cannot propagate by themselves and need man's guiding hand.

However, in the forests of Southeast Asia, where the banana was born, there are two extant exemplars of very seedy bananas with pits so hard that they can easily chip the molars of the absent-minded eater. These types are wild, their seeds produced through pollination. Botanists suggest that originally all bananas were like these seedy varieties, but that the evolutionary laws of chance produced bananas with fewer seeds and ultimately some with no seeds. Primitive peoples, who already used parts of the plant such as the male buds and the pseudo stem for food, recognized a good banana when they saw it and propagated the species that gave more for their efforts. Therefore, the seedless banana that we know is a fairly recent commodity, at the most ten to fifteen thousand years old.

The latest botanical classifications show the banana—and its cooking cousin, the plantain—as a member of the Musaceae family in the order of the Scitamineae. Like any good child of the tropics, however, the banana has a host of relatives near and far. One of its most distinguished relations is the household orchid, of the family of Lowiaceae. Other well-known relatives include the garden lily, the arrowroot, and the palm.

The Musaceae are split into two genera, the Ensete and the Musa proper. The Ensete plants are natives of Ethiopia and do not produce edible fruit; their bearings are so seedy that tribesmen use the pits to make necklaces. Ensete are considered monocarpic herbs, that is, plants that flower and die, such as the Mexican agave or century plant, whose sap is used to make tequila.

The Musa branch is split into four sections: Australimusa, Callimusa, Eumusa, and Rhodochlamys. The edible varieties arise from the Eumusa subsection; the others are known only as ornamental plants, except for Australimusa, out of which comes *Musa textilis*, or abaca, from whose filaments Manila hemp is made.

Originally, the further classification of the banana was simple. Bananas were *Musa sapientum* and plantains *Musa paradisiaca* (there being no genetic distinction between the two). This appellation was given by Swedish botanist Carolus Linnaeus. Today experts affirm that to continue classifying bananas after the Eumusa genus is fruitless. Fewer than half the banana cultivars in the world have been classified—botanically, we know of the existence of around 150 varieties, believed to be only half the varieties in the world. And no Latin name could possibly encompass all the Eumusa cultivars because they all vary genetically. Therefore, for instance, it's a botanical *faux pas* to write "*Musa sapientum*, var. Gros Michel," since *sapientum* refers to one particular type of banana, the silk fig.

Linnaeus's choice of the word *musa* for the banana had two possible origins. One was its usage among Arabs and Hindus—*mouz, moz, mawz,* or *moaz.* The other was by reference to Antoninus Musa, personal physician of the Roman emperor Augustus Caesar. Some philologists argue the obvious—the name is dedicated to the Muses—but this inspiration has not been confirmed.

The surnames for the banana, *sapientum* and *paradisiaca*, were derived from the two most famous banana legends. *Sapientum* was taken from Pliny's account of how the banana nurtured the sages of India; *paradisiaca*, from the Muslim tradition that blames the banana for Adam and Eve's exile from Eden.

While the classification of the banana can sometimes cause complications, the growth process of the plant itself is simple. Within a few days after planting, the rhizome issues a mass of cordlike roots. Some of these spread out laterally close to the surface for 10 or more feet, while others burrow vertically. Rootlets with root hairs for absorbing nutrients then run out from the main roots.

Leaves sprout from the eyes of the rhizome three or four weeks later. The first leaves grow out tightly furled into a spear shape that opens within a few days. New growths shoot straight out from the center of the leaves, rapidly compacting to form the pseudo stem.

During the first weeks bananas grow amazingly fast. Some species develop so quickly—the Latundan of the Philippines spurts out up to 20.5 centimeters, or about 8 inches, every twenty-four hours—that with a little patience you can see the plant sprout.

In a healthy specimen a new leaf emerges from the center by the time the previous one has opened completely, so that generally a plant produces about one new leaf per week. When the leaves are fully grown, they range in size from 3 to 12 feet long and in number from eight to fourteen, depending on the variety. The leaves are bright green and displayed vertically, creating a palmlike effect. They are very fragile, and splitting is extremely common, a trait botanists see as natural, analogous to the palm, whose single leaves are torn by winds into a tattered spray.

When grown, the plant ranges from 4 to 40 feet. Nine to twelve months after planting, the banana attains 90 percent of its height and is ready to bear fruit. The pseudo stem then contracts to half its previous width, and immediately afterward the flower bud, or inflorescence, starts its journey through the heart of the pseudo stem outward to the crown of the plant. The blossom matures into fruit within three to five months.

The flower bud, which grows at the end of a long stalk, is

covered by maroon-colored bracts, or sheaves, of remarkably rubbery texture. Initially, the bud swells in size and then ordinarily bends over and down, looking similar to an ear of corn in its husk. A few days later the bracts peel off, revealing the flowers and tiny fruit.

The flowers are arranged in clusters, borne spirally on the lone stalk. In each spiral there are twelve to twenty flowers. The basal, or top, flowers on the stalk are female, and the distal, or bottom, ones are male. The male flowers normally drop off, so the stalk has rows of bananas growing on top while the stalk below remains bare. In some varieties the stalk ends in an ovoid bud, or "blossom," made up of the remaining bracts enclosing unopened male flowers. In countries in the Far East that bud is often consumed as a vegetable, along with the flowers.

The complex of female flowers, having developed into fruits, is then called a bunch; each row of fruit is a hand, and each fruit a finger. The banana produces only one bunch, after which the plant withers and dies, while at its foot new shoots, or suckers, surge up from the rhizome to continue the phoenix-like life cycle of the banana.

In the commercial banana plant, the bunch weighs 50 to 60 pounds and hangs down because of the force of gravity. This is solely a characteristic of those cultivars chosen by companies to minimize damage to the fruit during packing. In nature many types of bananas grow their bunches differently.

Some cultivars grow a horizontal bunch; others, perpendicular; still others, at an angle from the ground. The shape of those bananas is different from that of the curvy berry found in fruit bowls all over America. Therefore we can now answer that old question, "Have you ever seen a straight banana?" The answer is yes, provided the banana comes from the top portion of a bunch that grew horizontally. The bottom portion of that bunch, however, will produce bananas shaped like semicircles, sometimes with an angle of 180 degrees between the tips.

How to
Grow
Your Own
Banana

OU need not live in Florida or Hawaii to boast of your own banana plant. The Sunbelt of the country—the South, the Southeast, and the Southwest up to 36 degrees latitude, or roughly the Mason-Dixon Line —is potential banana land. The banana, in spite of its preference for balmy tropics, adapts willingly to occasional temperatures as low as 26 degrees Fahrenheit.

The first step to successful cultivation is careful selection of the species. In general, you should pick one of the Cavendish varieties because of their adaptability to cool climates. Known under many names— Johnson in the Canary Islands, Camayenne in the French West Indies, Enano in South America—the Cavendish constitutes the basis of the banana trade of subtropical areas like Israel, South Africa, and the Canary Islands.

43

Cavendish comes in two sizes, Dwarf and Giant; the name refers to the height of the plant, not the measure of the fruit. The smaller of the species will grow under optimum conditions to 6 to 10 feet, while the giant surges to 12 to 20 feet. Both bear broad leaves on short stems, and their produce resembles the commercial banana, although somewhat shorter.

If you are among the fortunate few living in a frost-free climate year round, you can plant any one of the banana cultivars available. You may choose the apple banana plant, which soars to at least 20 feet; the Cuban red (no connection to the present Castro regime), with a claret-colored fruit; or even the famous Big Mike or Gros Michel, which yields large, plump bananas.

For a true banana plant, insist on a variety with the proper name of Musa; other types, such as the Ensete, while useful for garden ornamentation, produce inedible fruit. The exception to the Musa rule is the *Musa basjoo*, a Japanese genus grown for textile purposes.

Although the banana thrives in sandy loams, no set rules apply for soil quality or pH. However, never forget to drench the plant daily. Bananas require a minimum equivalent of 4 inches of rainfall a month; that means the more water the banana gets, the better it feels and grows.

The only other two requirements of the banana plant are intense sunlight and as much warmth as possible. You should plant the banana in the south or southeast corner of the garden to give it the most sunshine—as well as to shelter it from tearing winds.

The seedless banana reproduces by vegetative propagation, that is, from a bit of the rhizome or from a sucker. If you buy the shoot or sucker at a nursery, examine it carefully. Leaves from a healthy rhizome appear broad and whole, without ratty tears. Likewise, a young shoot ready for planting measures at least 3 to 4 feet in height. Prices for shoots range widely, but you can expect to pay from three to ten dollars for a healthy sucker.

If you can't find a nursery with bananas but you know of someone with a plant, ask for a cutting. Choose a sucker far away from the parent because it will have more fully developed roots. With a spade, dig in and cut firmly through the rhizome, making sure not to tear the rootlets when removing it.

If you are growing the plant outdoors, you should plant in the spring. If you use a greenhouse for the early stages, start setting in late winter, around February, so that the banana can take full advantage of the warm days of the summer, when it will sprout to full size.

In the garden dig a hole 30 by 36 inches for the Dwarf Cavendish or 40 by 60 inches for the larger varieties and excavate to a depth of about 18 to 30 inches. If the soil has a slight tendency to compact, deposit some sand or peat, as well as compost or manure, in the pit. If you set more than one stool, try to keep them a distance of 6 to 12 feet apart to allow room for lateral root expansion. Cover the base of the plant with at least 1 foot of earth, press down firmly, and water liberally. Again, remember the banana craves water. Check the site every day, particularly during hot summer days. If the ground feels dry to the touch, irrigate.

A heavy mulch cover around the plant helps retard weed growth, conserves moisture, and aids mineral nutrition. But bear in mind that excessive mulching may encourage superficial rooting and pose a threat to your banana. If you don't mulch, follow a regime of careful weeding during the first few crucial months; otherwise, the vegetation cover will steal the plant's food.

Fertilization, although not essential, speeds up the growth and improves the general health of the plant. The banana quickly depletes potash and nitrogen from the soil; these must be replenished during the first six months for maximum benefits. A quarter-pound of fertilizer with a high content of these minerals applied monthly in a circle 3 feet around the plant usually suffices. Follow the application with heavy watering.

Pruning is a matter of choice, but for a banana harvest you

must remove all but one or two of the suckers during the first stage of growth. Later, upon maturity, three or four may be spared their lives. The others should be removed by cutting in deeply with a spade and slicing off their share of the rhizome. Allow one sucker for each four months of plant life.

Banana plants are blessed with immunity to most diseases, save two—Panama and leaf spot, or Sigatoka. The first disease manifests itself by the splitting of the stem, blotchy yellow leaves, and the rotting of the rhizome. Panama cannot be cured, although it can be controlled by submerging the entire plant in water for at least six months. Most backyard growers might rather burn the plant.

Microscopic spores floating in the air cause Sigatoka disease, creating faint yellowish spots that later become streaks parallel to the ridge of the leaf and finally turn gray, burning off parts of the leaf. The infection affects mostly young leaves and reduces the chances of the plant's bearing fruit. Sigatoka can be controlled with a spray of a copper solution that inhibits the growth of the spores.

Pests, on the other hand, do very little damage to banana plants and fruit.

The banana plant can reach full size within twelve to eighteen months in the coastal areas of the Sunbelt; other zones may require eighteen to twenty-four months. A high incidence of frosts and low temperatures may further retard the growth of the plant.

Perhaps the best remedy for frost is artificial heat, which is a procedure used in growing wine grapes. Australian planters initiated this method; they maintain heaps of sawdust smoldering constantly for three months of the year. A wood fire, stoked properly, gives off the necessary warmth too. You should also consider planting the banana next to a vent from a dryer or a heater.

If frost damages your plant, however, don't despair, for it probably won't die. The freeze will make the leaves drop, but you can protect the stalk by mulching and covering it with burlap. This is particularly advisable in regions with long winters. If frosts are relatively uncommon, don't touch the plant; simply observe it periodically for new growth. When the first leaf shows, shear off the damaged leaves, pruning down and at an angle. You should leave just the stalk or stem and the new growth, and within a few months the plant will replenish itself.

And fruit? Well, fruit will come but will require some patience: in the Sunbelt, bearing may come about eighteen months to two years after sprouting. Barring killing frosts, however, you will ultimately be rewarded with your own bunch of bananas; in the meantime, there are graceful fronds to sit under while sipping a gin and tonic and dreaming of tropical romance.

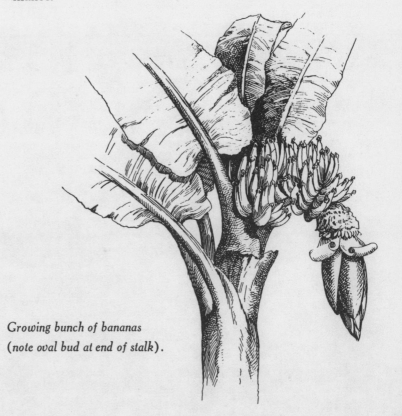

Growing bunch of bananas
(*note oval bud at end of stalk*).

Tricky
Banana

AN, in his search for improvement on Mother Nature, has finally figured out how to slice a banana without peeling it. This stunning discovery was brought off by two enterprising New Yorkers, Lee Eisenberg and Tom Farrell, whose patriotism compelled them to find an American match for the Japanese square egg:

1. Take a ripe, brown- or black-flecked banana and an extra-fine needle long enough to reach from one side of the banana to the other. Thread the needle with about 2 feet of synthetic or fine silk thread.

2. Dip the needle in lemon or lime juice to prevent staining and pierce the skin of the banana at any point along one of its ridges.

3. Pass the needle just far enough under the skin so as not to leave any tracks and draw it out at the next ridge.

4. Pull the needle and thread through the hole, leaving the end of the thread protruding from the starting point.

5. Dip the needle in juice again and insert into the newly made hole in the second ridge. (It's important to go back through the *same* hole.) Pierce through as before to the next ridge. Continue until all ridges have been pierced and needle and thread are brought out through the original entry point. The banana in cross section would look like this:

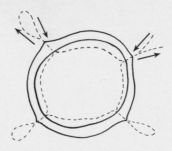

6. Grab both ends of the thread and pull. The pulp, which has been girthed by the thread, will be sliced through at this point. Do not pull too sharply or you may create too great an opening in the skin.

7. Repeat the process as many times as desired. The banana will be sliced within its casing, and only the sharpest eyes will detect the trick.

8. Peel and enjoy!

A View
of the
Plantation

 BANANA plantation is an agricultural factory where the great provider, nature, manufactures the product that banana companies keep forwarding to your supermarket shelves.

Throughout the year, banana economists calibrate consumption and demand, signaling to computers the orders that dictate plantation production. All actions and decisions are taken with one eye on the clock and another on the calendar, from the time rhizomes are planted to the day boxes arrive at the wholesale market.

Planting is the only operation accomplished with ease (nature can be primed but not hurried). During the time of plantation site selection, agronomists juggle such factors as fertility of the soil, access to sources of water, and costs of materials needed. Once the terrain is

51

chosen, it is cleared of previous growth; if woodland, all underbrush is chopped down and left to rot in the ground; if an agricultural field, the vegetation is cut and then burned.

Workers then line the land with rows of sticks in a grid pattern, the intersections being the planting spots. The planting layout is on a square, and spacing between plants varies according to the variety grown; rhizomes or young plants are set down in holes 18 to 30 inches deep that will accommodate the long burrowing roots. Long lines of ditches are dug, crisscrossing the fields, to drain water from marshes and channel tropical rainfalls.

Plantations on the Pacific coast of Central and Latin America receive less than the minimum banana requirement of 4 inches of rainfall per month. To compensate, banana companies have set up towers for "overhead irrigation"; revolving nozzles atop these 26-foot structures spray water in circles for 3.3 acres.

As the suckers emerge during the months after planting, the site is periodically cleaned to prevent weeds from sapping the nutrients needed by the banana. Chemicals are sprayed in some spots, but the greater portion of the weeding is still carried out by the traditional method, with a machete. Plastic bags are also hung around developing bunches to protect them from feeding birds and insects, as well as inclement weather.

Within ten to twelve months the plants are ready for harvesting, and the entire machinery of distribution gears up. Ordinarily the fields of a plantation have been planted at different times to assure steady, constant production. Because only a very short span elapses between the time of harvest and the date of delivery abroad, every minute is accounted for. Computers match up the orders coming in from the main distribution centers in the consuming countries—which in the United States are Los Angeles, New York, and New Orleans—with the corresponding fields, indexed and processed down to the last bunch.

Every day at dawn bands of men set out to bring in the

green but fully developed bunches. They work in pairs, a cutter and a carrier. The cutter chops the stem of the plant, which slowly bends down. The cutter places the bunch on the shoulder of the carrier and, with one slash of his machete, slices the bunch from the stalk. While the carrier drags away the bunch, the cutter levels the stalk to a stump that will grow and bloom the following year. Decades ago the bunches were stacked in carts and drayed to dock by mules. Today the fruit is hauled to the packing plant by tractors, trucks, or even wire-cable conveyor systems strung along jungle paths.

At the plant, workers place the bunches on a monorail and push them to the washing station, discarding grossly substandard fruit. At the showers, the stems are washed with a fine spray of water to remove traces of insecticides, as well as dirt and insects. Afterward stems are again inspected and impaired fruit is rejected.

The bunches then pass to the cutters; a crew removes the flower tops from the banana fingers and weighs and places the bunches in the cutting line. Cutters reject undersized and undesired fruit; defective hands and fingers are pulverized and flushed away in a sluice. The bunches are cut into hands and stream down a tank of water to a selector at the opposite end of the tank. This washing also cools down the fruit and drains off the rubbery sap that flows from cut green bananas.

The selector is ordinarily a woman; companies seem to believe a feminine touch does not injure the easily bruised fruit. The selector fills up trays with 42 pounds of bananas, which after loss from evaporation will weigh about 40 pounds. She places several trays on a conveyor that carries them through a bath of fungicide. After drying, a packer fills up boxes with hands, fits in the lids, and forwards the boxes to a refrigerated room, where they remain stored until shipment. That is the last time human hands touch the banana until the day the retailer rips open the box. Each box bears a stamped number, so any irregularities can be traced to the plant, the packer, the selector, and even the farm where the fruit grew.

Railroad cars carry the boxes to port, where gantries, or conveyor belts with pockets, load the bananas into the refrigerated hold of a steamer. These ships have a capacity of over 5 million pounds; normally, they take on a full load in less than a day. The bananas are then shipped to their destination and handed over to the wholesaler, who ripens them and sends them to market.

In this manner banana companies deliver the fruit year round to your table with the smooth continuity of a Detroit shop. The entire procedure of harvest, shipment, and distribution is accomplished in ten to twenty days. In fact, bananas need only wheels to announce more clearly their status as mass-production items. There is a difference, though: there has never been any recall of bananas in the history of the trade, a record Detroit might well envy.

Where
the
Banana
Grows

I learned to drink my likker
Way down in Costa Rica—
'Taint nobody's business what I do . . .

HAT ditty was one of the favorite tunes of turn-of-the-century banana plantation workers in Costa Rica, the original "banana republic." The Caribbean coast of that bean-shaped nation became the first commercial banana land because it met the requirements of optimum banana production. Deep alluvial soils, often up to 20 feet thick with earth cover, guaranteed bumper crops without fertilization, while heavy rainfall assured the plants their daily water fix without the expense of irrigation. In addition, an average monthly temperature of 80 degrees Fahrenheit hastened the growth and bloom of the banana. A deep harbor and a serviceable railroad line made the transport of implements to plantations as convenient as the transport of fruit out to the buying world. Moreover, there was an accommodating government willing to grant special con-

cessions, such as elimination of duties on imported goods. Finally, when the first banana lands were exhausted (as most are, within ten to twenty years), there was always more to grab, virgin land that previously served only to breed snakes, pigs, and pestilence.

When time came for expansion, banana companies searched for land that combined those qualities. They found it in Honduras, Guatemala, Nicaragua, Panama, Colombia, and Ecuador. These are still the world's top banana countries, producing two-thirds of the commercial global crop. (The Philippines, a late arrival, by 1978 had captured a 10 percent share of the world's exports.)

But bananas don't require Latin enticements to grow. They also thrive in all countries within the 30-degree lines of latitude. This portion of the earth includes the rest of South America, except Chile and Argentina; Africa; the Arabian Peninsula; Iraq, Iran, Pakistan, and the remainder of Asia, excluding Korea; and all the islands, reefs, and atolls of the Pacific—in sum, over one hundred countries. However, this is just an overview; in some instances, as in Israel, bananas are found as a commercial crop outside those geographical limits.

On the other hand, most nations do not cultivate the banana for commercial purposes. In some countries, such as Italy, the banana is an odd ornamental plant, looking tattered and incongruous amid imperial ruins; in Zambia or South Africa, it is just a local foodstuff.

A number of conditions besides favorable climate must be satisfied before bananas become an export crop. The first is an internal commercial structure for cultivation and distribution of the fruit—roads or railroads, deep-water ports, advanced communications systems, and so forth. This requirement in most instances necessitates a powerful foreign company, as in Central America, or a well-subsidized government concern, as in Ecuador.

Another prerequisite is the availability of large tracts of land that is undeveloped because of a low population density by comparison to the total national territory, as in Costa Rica, where the majority of the population lives in the central highlands and only a few thousand farmers reside in the sparsely settled coastal areas.

But perhaps the most important condition is a world price that will turn a desirable profit for the growing country. Brazil, the world's largest grower of bananas, exports only a tiny fraction of its produce because the gain isn't worth the bother. Were the craving for bananas in consuming countries to rise to the level of Uganda, where natives eat over 9 pounds of bananas a day, demand would stimulate a price hike that would encourage countries such as Mexico, India, and Indonesia to join the banana export club. That prospect, though, is probably just another case of bananas, or unmitigated nonsense.

Bananania

F ALL the bananas imported into the United States were placed end to end, the chain would make a loop twice the distance from the earth to the moon.

In Council Bluffs, Iowa, it's against the law to sell bananas without warning the buyer on the dangers of casting the peels on the sidewalk.

The banana is the third largest export of Iceland. The fruit is grown in greenhouses heated by water pumped up from volcanic underground springs.

Scene from The Gang's All Here, *with Carmen Miranda and Alice Faye, directed by Busby Berkeley. Courtesy 20th Century—Fox.*

Banana fibers from the stalk or pseudo stem of the banana plant can be woven into an attractive silklike fabric. Garments made of such cloth are worn in the Solomon Islands, the Caroline Islands, China, and Malagasy. Hats are woven from the fabric in Japan and Indochina.

In Uganda new mothers are given the sap of the banana plant to drink because it is believed to help increase lactation.

The founder of United Fruit Company, Minor C. Keith, disliked bananas.

Arabs were raising bananas in southern Spain in the tenth century.

''Throughout Equatorial Africa, there is a drink prepared in the following fashion: green bananas are buried in holes lined with straw; they are left there to ripen for a week; then the peel is taken off; they are mashed, then placed in a container with water. Two days are required to make a drink that is much liked, even by Europeans.''

> Translated by the author from *Le Bananier* by Paul Hubert, Paris, 1907

In 1977 Geoffrey Thrippleton Man, last captain of the Cunard Line's famous cruise ship *Queen Mary*, was serving as second mate in a banana boat.

In 1974 two explorers trekking across the Chilean desert in a Citroën 2CV noticed that their car's sump oil had drained out.

Miles away from civilization, they grew desperate searching for a substitute. They ransacked their supplies and came upon a hand of bananas. They peeled the bananas, mashed them, and placed the mash in the sump. The bananas acted as a lubricant and the explorers managed to avoid death in the dunes.

Italians have a favorite phrase to indicate the impossible: "*Ma va a drizzar banane*," or "Go straighten out bananas."

Every year the border towns of Fulton, Kentucky, and South Fulton, Tennessee, hold the world's only International Banana Festival. Thousands of visitors attend the feast, consuming over 10,000 tons of bananas and a 1-ton pudding served with paddles. Banana-producing countries send representatives to lecture on the history of their republics and artists to demonstrate native crafts. The festival was conceived in 1963 to honor the history of Fulton and South Fulton as the "banana crossroads of the world," a designation that stems from the early days of the banana trade, when the towns were a central checkpoint for refrigerated railroad cars en route to the Midwest from New Orleans.

The natives of the province of Buganda, Uganda, are the greatest banana eaters in the world. They consume an average of about 9 pounds of bananas a day. In fact, according to Ugandan law, a basic food ration for a working man consists of 8 pounds of *peeled* bananas a day.

Herbert Talbot, a carpenter from the Isle of Wight, England, was awarded £7,500 in 1972 for a road crash injury that deprived him of his sense of smell and of taste. Mr. Talbot had pleaded that "the only thing I can really taste is bananas and hot custard."

Scene from The Gang's All Here, *with Carmen Miranda and Alice Faye, directed by Busby Berkeley. Courtesy 20th Century–Fox.*

At the turn of the century banana coffee was a popular substitute for real coffee.

In 1923 the Westman Publishing Company, publishers of the sheet music for Handel's *Messiah*, sued the writers of "Yes, We Have No Bananas (Today)," claiming that the melody had been stolen from the *Messiah*. Westman won the suit.

According to the *Guinness Book of World Records*, the longest banana split ever concocted was 1 mile, 99 yards long. A total of 11,333 bananas, 34,000 scoops of ice cream, 260 Imperial gallons of topping, 160 pounds of chopped nuts, and 100 gallons of whipped cream were needed for this Brobdingnagian delight. The dish was made at the annual fete of the Cleveland State High School in Queensland ("Bananaland"), Australia, on November 20, 1976.

The first, and so far only, registered use of bananas as a political weapon took place in May 1977 at Coacalco, Mexico, where four thousand enraged citizens captured the town mayor and forced him to eat 12 pounds of bananas as punishment. The townspeople were protesting the shooting of a workman and had already stoned the local police commander and his deputy. Mayor José Ramón del Cueto promptly signed his resignation after he crammed his course of banana power.

Bananas are the main product of the biblical town of Jericho, Israel.

Years ago natives of the province of Queensland, Australia, were called "Bananalanders" because of the many banana plantations in that area.

The heart of the banana stalk, or pseudo stem, was once a highly prized delicacy in India—about half a ton a day was sold in Calcutta alone during the 1890s.

A banana mask can do wonders for tired faces. Because the fruit contains a great deal of potassium, which is quickly absorbed by the skin, the mask tightens the skin while relaxing the facial muscles. You can apply thinly sliced bananas directly to the face or make a cream of a third of a banana mashed and mixed with a tablespoon of honey. Leave it on your face for about twenty minutes and rinse off. Your skin will feel soft and fresh for about two days.

The uplands of Uganda and Tanzania, to the west and north of Lake Victoria, are the most heavily laden banana areas in the world.

In New Caledonia bananas are ripened by burying them in mud packs for several days.

The ash of the banana plant is an alkaline-rich caustic that can be used safely for washing clothes.

The first shipment of bananas to the United States was reputedly made in 1690. The bananas, which had been raised in Panama, were taken to Salem, Massachusetts. The Pilgrim fathers boiled them for a pork dinner and many diners later complained that the newfangled fruit tasted like soap.

"Feeding the Monkey" is an amusing party game. To play, two persons who are blindfolded sit or kneel on cushions about 3 feet apart. Each one is given a peeled banana and told to "feed the monkey"—that is, place the banana in a position where the other player can bite it. The one who eats the most of his or her banana within three minutes wins. Fans of the game say they break up in laughter when they see the players miss each other's mouths and hit instead the face, the neck, or other parts of the body.

An acre of bananas produces a minimum of 18,000 pounds (9 tons) of bananas a year.

A whiskey made of fermented bananas was awarded a gold medal at a St. Louis exposition at the turn of the century. The whiskey had been aged only six months in the barrel and its purity confirmed through chemical analysis by the U.S. Department of Agriculture. Imbibers at the time said its taste resembled that of Canadian Club.

In the West Indian island of Grenada towering plantain plants are used to shade cacao and nutmeg bushes.

The *fe'i* banana of Tahiti is one of the few extant varieties of wild bananas. The fruit grows in the island's hill forest. Every season professional *fe'i* hunters follow ancient trails up the hills to gather the bananas and return with loads of bunches weighing 150 pounds.

Tahitians have found many uses for the *fe'i*. Its green leaves serve as plates and as roof thatching. When dried, they are also employed for bedding, packing parcels, and even for making cigarette papers. In addition, strips from the midribs of the leaves are turned into thongs and lashings; finally, when carefully stripped, the leaves serve to make fans and mats.

The sap of the *fe'i* banana is reddish violet, and one time a resourceful missionary, short of ink, copied an entire Bible with bamboo pens dipped in the red juice.

Banana skins are ideal for shining tan shoes. Rub the inside of the peel against the shoe, let dry a few minutes, and wipe away the coating. Buff with a dry cloth. The tannic acid and coloring give a beautiful shine to your tans.

The common banana in the United States today owes its varietal name, Cavendish, to that of the first cultivator of the fruit in English soil, Spencer Compton Cavendish, eighth duke of Devonshire, in whose greenhouse the banana flowered in 1836.

A Cuban folk tale warns that drinking rum while eating bananas is poisonous. According to this legend, when British troops invaded Cuba in the eighteenth century, wily Cubans drove out the Redcoats by giving them primitive banana daiquiris. A similar story is told in Jamaica; however, there the islanders malign only one type, the claret or red banana, commonly called Mankiller.

In the banana plantations of Honduras workers tell stories of a simian creature named Sesimite or Oso Caballo that lives in the swamps. Some American technicians think he must be the southern cousin of the Pacific Northwest Bigfoot. The workers describe him as over 6 feet tall, with a rank, hairy body and almost human features, but they've never seen him eating a banana.

Michael Gallen, twenty-three, of Australia, holds the world

record for eating the greatest number of bananas in the shortest time. According to the *Guinness Book of World Records*, Gallen downed sixty-three bananas in ten minutes on October 11, 1972.

The tall Montecristo variety of banana is used in Puerto Rico to shade fragile coffee bushes.

"If only the virtues of banana flour were publicly known, it is not to be doubted but it would be largely consumed in Europe. For infants, persons of delicate digestion, dyspeptics and those suffering from temporary derangement of the stomach, the flour properly prepared would be of universal demand. During my two attacks of gastritis, a light gruel of this, mixed with milk, was the only matter that could be digested."

> **Sir Henry Morton Stanley**
> *In Darkest Africa*
> (New York: Scribners, 1890)

Chinese pig farmers in Malaysia believe cooked banana stems served as fodder prevent kidney worm disease (*Stephanurus dentatus*) in their pigs.

"We do not want to imitate nature. We do not want to re-create, we want to create. We want to create as the banana tree creates bananas."

> **Hans Arp**
> in *Concrete Art*

In 1914 the retail price of bananas in most cities in the United States was fourteen cents a dozen.

"A green banana is as good as a yellow banana."

André Breton

Many technical books on bananas from the turn of the century recommended distilling brandy from ripe bananas. The proportions given specified that 60 bananas were needed to make 2.5 liters of brandy.

"The bankruptcy of ideas having destroyed the concept of humanity to its very innermost strata, the instincts and hereditary backgrounds are now emerging pathologically. Since no art, politics or religious faith seems adequate to dam this torrent, there remain only the bleeding pose and the banana."

Hugo Ball
in *Dada Fragments*

The
Political
Banana

DON'T know that I deserve credit for the shape of the country I produced in Czechoslovakia. It looks something like a banana."

Former CIA director Allen Dulles, speaking of his role in the post–World War I Boundary Commission

"Oh, the delicious fruits that we have here and in Syria! Orange gardens miles in extent, citrons, pomegranates; but the most delicious thing in the world is a banana, which is richer than a pineapple."

Benjamin Disraeli (1804–1881), British Prime Minister, in a letter to his sister from Cairo, Egypt

71

Argentine-born revolutionary Ernesto "Che" Guevara, during a stay in Guatemala in the 1950s, applied for a job with United Fruit as a doctor on a banana plantation. He was turned down.

"When I was Cabinet-making I had sandwiches as my nourishment. What a better job I might have made of it on bananas!"

British Prime Minister David Lloyd George (1863–1945), reflecting on his political career

"Go pick your bananas and we'll run the Canal!"

New Hampshire governor Meldrim Thomson in 1975, upon hearing Panamanians wanted a say on Canal operations

Angel Castro, father of the Cuban ruler Fidel Castro, was an employee of the United Fruit Company. When young, Fidel tended the bananas on his father's 10,000-acre estate in eastern Cuba.

"At breakfast time peel a banana. Lay it on a plate of ample size and ladle over it thick orange marmalade. Eat with a spoon, munching at the same time hot buttered toast. Sip coffee."

Breakfast recipe of Edward VIII, Duke of Windsor (1894–1972)

The
Fans
of the
Banana

ANY are the fans of the banana; from boardroom to ghetto shanty, from capitols to lettuce fields, on the highest mountaintops, in the deepest valleys, in the plainest plains, their names are legion. Here are the stories of two such fans, dedicated to the proposition that the road to happiness and fulfillment is lined with bananas.

One fan so surrendered herself to the ecstasy of the fruit that she changed her name. Today she is known solely as Anna Banana, and her love is pure, for although she has reaped no earthly rewards, she stands fast in her faith. Anna spreads her message from a modest home in San Francisco's Noe Valley. She is in constant contact with her correspondents, the banana votaries of the world, who endow her with representations of the object of their devotion. Her walls are lined with banana objects, dolls, pillows, plaster casts, and myriad modest drawings, while in her files she stores the story of the banana as recounted in articles and books in the major Western languages.

A Dada artist, Anna Banana founded the School of Bananology several years ago to proselytize "banana consciousness." The school, a sectarian institution, awards diplomas to those who approach it thirsting for the truth of the fruit. No disciplines are imposed, but converts are requested to contribute a bit of banana lore before receiving a Master of Bananology degree from the Order of the Banana.

One of the school's extracurricular activities is the Banana Olympics, held annually at San Francisco's waterfront Embarcadero Plaza. Each year hundreds of contestants vie for the honor of competing in the ceremonies. In 1978 Anna officiated on April Fool's Day; over four thousand believers attended, many participating in the diverse celebrations—the Belly to Belly Banana Race; the Overhand Banana Toss (a banana javelin throw); the Banana Song Contest. The winners received a banana on a ribbon as first prize.

Anna, the founder of *Vile* magazine, a Dadaist journal patterned after *Life* and *Look*, has received some academic support for her efforts, but the cost of casting the word of the banana, of printing and distributing the journal, comes from her own pocket. To make ends meet, she hires herself out as a singing banana to an outfit called Rent-a-Yenta; dressed in a banana costume, she taps and croons and recites banana history to hosts on special occasions.

If Anna is the Peter of the banana, then Ken Bannister is the Paul. By day Bannister is president of a photographic equipment concern at Altadena, a small town near Los Angeles. But by night he dons the vestments of top banana at the International Banana Club, an incorporated hierarchical organization with over five thousand members scattered like so many peels around the world. The purpose of his brotherhood is simple: "to influence more people to smile more often." To that purpose, Bannister begat the club's own lingo, bulletin, and assorted paraphernalia, all registered under the trademark Banana Club®. Members are charged annual dues of two dollars or contribute ten dollars for lifetime membership, which entitles them to a membership card, a banana title, a twenty-four-karat-gold-plated lapel pin, two bumper stickers, a membership certificate, and all the editions of the *Woddis News*.

IBC members are ministers of the peel, each baptized with a banana name upon ordination. There is the radio announcer in Hawaii dubbed the Banana Voice; the Federal Aviation Administration official hailed Top Flight Banana; and even a Banana Palestrina, a famous but unidentified composer who wrote the club's paean. Lower members are divided according to banana merits earned through Banana Activities; they are classified from Right Banana Hand and Finger Director to Master and Doctor of Bananistry. All Banana Activities are prefaced by the sacramental words "Woddis" and "Ja-wow." These terms are akin to the greeting of "Peace" and replace wilted expressions such as "Hello," "Good-bye," and "Far out." The club also has a Museum of Bananistry, a collection of banana-related items "in good taste," as befits the virtuous banana.

Anna Banana
1183 Church Street
San Francisco, California 94114

International Banana Club
2524 North El Molino Avenue
Altadena, California 91001

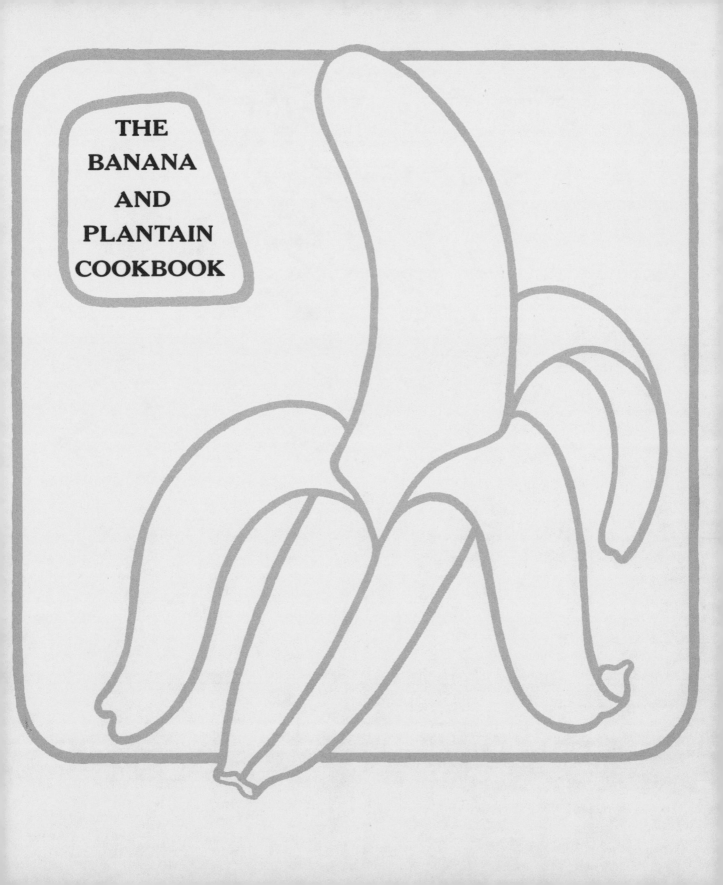

THE
BANANA
AND
PLANTAIN
COOKBOOK

On
Cooking
Bananas

O NLY a small portion, perhaps 5 percent, of the billions of pounds of bananas sold in the United States today are cooked. That is a shame, for few foods can match the rich, savory flavor of cooked bananas.

Years ago, when they were more exotic, bananas were commonly served roasted or broiled with meats. One such dish, fashionable at the turn of the century, was offered by Delmonico's, the restaurant that catered to the industrious appetites of Diamond Jim Brady and J. P. Morgan. Called Steak Stanley, the entrée consisted of sirloin or porterhouse steak, served rare, accompanied by broiled bananas in horseradish sauce.

Unfortunately, today few Americans consider bananas for cooking as vegetables. The following recipes, culled from the many foreign cuisines in which bananas and plaintains are staples, show what you probably have been missing. These recipes generally call for ingredients readily available in most stores. Those few ingredients of tropical origin that might prove unfamiliar can be found in any Latin American market.

SPICES

Perhaps the most unusual condiment used in the following dishes is the seeds of the annatto tree, or *achiote*. Produced by a tropical tree, these seeds are known as a food coloring throughout Latin America; their dyeing properties were discovered by Caribbean Indians, who daubed their bodies with *achiote* during certain ceremonies. An orange-red pulp surrounds the seeds; when fried, it turns the cooking oil or lard a deep orange color. The seeds should be scooped out before the oil turns a lighter shade, which indicates that the heat is breaking up the molecules that render the color and mellow taste. A good proportion for *achiote* oil is 4 tablespoons of seeds to ½ cup oil or lard.

In some countries the annatto pulp is ground and sold in small cans under the name *bixa* or *bixol*. A pinch of this powder dissolved in liquid will serve the same purpose as the seeds. Saffron may be substituted if no *achiote* is available.

The other herb that your supermarket might not stock is known as *cilantro* in Mexico and *culantro* in the Caribbean; we call it coriander. But while in this country we ordinarily use only the seeds, in other cuisines the entire herb is consumed. A relative of common parsley, *cilantro* looks similar to Italian parsley; it is the secret of a great *seviche*, or raw pickled fish. Chinese stores in many cities sell *cilantro* under the name Chinese parsley. Some people, however, feel the flavor is too pronounced, so moderation should be followed.

VEGETABLES

One of the traits of tropical diets is the ample consumption of starchy produce. In some cases these vegetables resemble North American vegetables, as happens with *calabaza*. Also called West Indian pumpkin, *calabaza* bears the shape of a large acorn squash. Its seedy center should always be discarded. *Calabaza* quickly dissolves in boiling liquids and is therefore added to soups and stews as a thickener. Its taste approximates that of Hubbard squash.

Other vegetables frequently found in tropical meals are the *malanga*, the *yautia*, and the *yuca*. *Yuca* is an edible tuber also known as cassava, used to make tapioca. Caribbean Indians fashioned a sort of unleavened bread from cassava that I remember sampling during a fifth-grade history lesson. The cake was oblong, about 3 feet by 2, was toasted a golden color, and had a dry, crispy taste like that of a saltless cracker. The *yuca* root, when peeled and boiled, is soft except for the woody center filament that must always be removed; the tuber's bland flavor offsets spicy sauces very well.

Malanga and *yautia* are tubers related to each other and to *yuca*. The first shows a white color when peeled, and the latter a faint yellow tint. *Malanga* is also called white *yautia* by Puerto Ricans; both these vegetables are drier and mealier in texture than *yuca*.

Some dishes in this book also call for *garbanzos*, or chickpeas. Members of the pea family, *garbanzos* are part of the Spanish cooking legacy in Latin America; many Castilian stews substitute chickpeas for beans. *Garbanzos* are also consumed in Italy and can be found in any Italian delicatessen, canned in water, under the name *ceci*.

MEATS

American cuts of beef are employed in the recipes here. However, two pork by-products mentioned may not be widely known. One of these is *chorizo*, or Spanish sausage. There are several types of *chorizos*, depending on the country of origin, but the most widely available are the Mexican and the true Spanish style.

Mexican *chorizo* contains a great deal of *achiote*, salt, and spices and crumbles rather easily when cooked. Spanish-type *chorizo*, on the other hand, presents a less formidable attack to unseasoned taste buds; it's also packed in a stronger casing, so the sausage remains whole even after prolonged cooking. For most recipes here I advise using the Spanish type, if available. If not, any moderately spicy Italian sausage will serve reasonably well.

Chicharrones, another widely popular food of Latin America, are pieces of the rind of the pork fatback, fried until the lard is tried and the skin is crisp. They are called cracklings in soul cookery. *Chicharrones*, dense and chewy, are always supposed to contain a morsel of meat. The aerated, synthetic bar snacks sold in plastic bags don't resemble them in the least.

EQUIPMENT

All these recipes can be made easily with the usual range of implements in the American kitchen. In fact, with the latest fad for food processors and other electric wonders that remove much tediousness from cooking, most of these dishes should be easy indeed. One piece of equipment you might consider acquiring, should you lack it, is a thick-bottomed, heavy cast-aluminum or cast-iron casserole with a tight-fitting lid, often used for slow-cooking stews. An earthenware casserole will also serve very well if you place enough asbestos pads underneath so that direct heat won't crack it. A Dutch oven will do quite nicely too.

HOW TO PEEL AND SEED A TOMATO

The easiest procedure for peeling a tomato is to drop it into boiling water. When the skin cracks, remove and pare. The length of time the tomato needs to be in the water depends on the thickness of the peel. Tender tomatoes can be peeled after 30 seconds, others require 2- to 3-minute dunkings. Once you have peeled the tomato, squeeze it onto a plate. The juice and seeds will ooze out.

HOW TO ROAST A GREEN PEPPER

Roasting gives the pepper a nutty, rich flavor. The original method, still followed in some parts of Europe, was to place the pepper under the embers of an open fire for about 5 minutes. Today the simplest way is to pierce the top of the pepper with a kitchen fork and place the pepper directly over a range burner, turning it until the entire skin turns black. (If you have an electric stove, broil the pepper instead.) Once it is roasted, you may let it cool for about 10 minutes, when it will peel easily. If pressed for time, you may scrape the skin off under running water.

Hors d'Oeuvres, Appetizers, and Accompaniments

BANANA BACON BITS

5 green-tipped bananas
½ pound bacon

Peel the bananas and slice into 2-inch rounds. Cut the bacon slices into 4-inch pieces and wrap each around a banana slice, fastening it with a toothpick. Fry, turning frequently, until the bacon is cooked. Remove, drain on paper towels, and serve hot.

About 20 bits

BANANA CHIPS

6 green-tipped bananas
2 cups oil
Sugar

Peel the bananas and slice into ¼-inch-thick chips. Heat the oil in a frying pan to approximately 375 degrees, or until a bread cube dropped in turns brown. Fry the chips a few at a time for about 5 minutes. Remove, drain, and dust with sugar. Use on cereal, yogurt, with sweet dips, etc.

Note: Do not sprinkle salt on fried bananas; it imparts a highly disagreeable flavor.

About 200 chips

BANANA DIP

1 medium, ripe banana
1 8-ounce package of cream cheese
1 tablespoon milk
2 tablespoons mayonnaise
1½ teaspoons curry powder
Salt and pepper

Mash the banana and combine it with cream cheese that has been softened with milk. Mix in the mayonnaise, curry powder, and salt and pepper to taste. Chill for 2 hours and serve with carrot sticks, celery stalks, banana chips, etc.

About 1 cup

CHIAPAS DIP (MEXICO)

2 ripe bananas
Juice of 2 limes
2 ripe avocados
¼ cup sour cream
¼ cup chopped walnuts
Sugar
Salt

Peel the bananas and mash with the juice of 1 lime. Peel the avocados, mash with the juice of the other lime, and combine with the bananas and sour cream. Add the walnuts and sugar and salt to taste and mix well. Chill. Serve with tortilla chips.

About 2½ cups

BANANA BEAN DIP (GUATEMALA)

1 tablespoon oil
2 *chorizo* sausages, skinned and chopped
1 onion, chopped
2 cloves garlic, minced
¼ cup Salsa Ranchera (see page 159)
½ cup tomato sauce
1 teaspoon sugar
Salt and pepper
2 cups cooked kidney beans
4 ripe bananas

In a frying pan, heat the oil and fry the sausage meat. Cook over medium heat for about 5 minutes. Remove the meat. In the sausage fat, fry the onion and garlic until soft. Add the *salsa*, tomato sauce, sugar, and salt and pepper to taste, and cook over medium heat for 5 minutes. Return the meat to the pan and add the beans and the bananas sliced in rounds.* Mix, cover, and simmer for about 5 minutes, or until the beans and bananas are warmed through. Serve with tortilla chips.

About 3½ cups

* Unless otherwise indicated, banana slices should be ½ inch thick.

CARIOCA BANANA (BRAZIL)

4 ripe bananas
¼ cup oil
1 large onion, chopped
¼ cup dry bread crumbs
½ teaspoon cayenne, or Tabasco sauce to taste
Salt

Peel the bananas and slice into 2-inch rounds. Fry in the oil until golden brown and remove. In the same oil, fry the onion until translucent. Meanwhile, mash the fried bananas with a fork. Add the cayenne or Tabasco to the fried onion, add salt to taste, and stir in the mashed bananas and bread crumbs. Cook for about 2 or 3 minutes, stirring all the time. Serve hot as a side dish.

Serves 4

ALCAPURRIAS (PUERTO RICO)

Filling:

1 medium onion, chopped
1 green pepper, chopped
2 cloves garlic, minced
2 tablespoons oil
1 ounce salt bacon, diced
   (optional)
½ pound ground pork
½ pound ground beef
1 medium tomato, peeled,
   seeded, and chopped
2 ounces boiled ham, chopped
1 tablespoon capers
2 tablespoons pimiento-stuffed
   olives, sliced
   Salt and pepper

Dough:

3 green plantains (or 5 green
   bananas)
1 pound *yautias*
¼ cup *achiote* oil
1 teaspoon salt
   Oil for frying

First prepare the filling by frying the onion, green pepper, and garlic in the oil until soft. Add the bacon, if used, and the pork and beef. Fry for about 5 minutes. Add the tomato, ham, capers, and olives and cook slowly for about 30 minutes, or until the pork is completely done. Season with salt and pepper to taste.

Meanwhile, peel the plantains and *yautias* and boil in salted water for 5 minutes. Remove, drain, and grate finely, then combine with the *achiote* oil and salt. Shape the mixture into 10 or 12 rectangles approximately 4 inches wide and 6 inches long. Place a tablespoon of the meat mixture in the center of each rectangle and roll it into a cylinder. Deep-fry in oil until brown.

10 to 12 alcapurrias

PLANTAIN CHIPS

4 green plantains
2 cups oil
   Salt

Peel the plantains and slice into ¼-inch-thick chips. Heat the oil to approximately 375 degrees and fry the chips for about 5 minutes, or until golden brown. Remove, drain on paper towels, and dust with salt. Serve with cold beer.

Serves 6

LIMBO CAKES (JAMAICA)

2 large, green plantains
4 cups salted water (optional)
2 cups oil

Bearing such names as *tostones* in Puerto Rico and *chatinos* in Cuba, limbo cakes are found throughout the Caribbean. They are served while warm and are always accompanied by a cool drink. This is without doubt the most popular way of eating plantains in this continent, and one of the simplest as well.

Peel the plantains and slice into 2-inch rounds. If desired, the slices may be soaked in salted water for 15 to 30 minutes to eliminate some of the starchy flavor of the plantain. Drain the slices and pat dry with paper towels. Fry at 375 degrees for about 5 minutes, turning over once. Make sure the slices do not brown but remain a golden color. Remove from the oil and drain. With a rolling pin or the flat side of a mallet on wax paper, or with the heel of the hand on a rolled-up paper bag, flatten each slice to ½-inch thickness. Return the slices to the oil and fry over high heat until brown and crisp. Remove, drain, and sprinkle with salt.

Serves 4

DOS RIOS PLANTAIN (JAMAICA)

2 medium, ripe plantains
   Salt and pepper
½ cup flour
½ cup bread crumbs
1 cup butter
½ cup brown sugar

Peel and cut the plantains into 1-inch-thick diagonal slices. Sprinkle with salt and pepper, then dip in flour and bread crumbs. Melt half the butter in a large skillet and add half the slices. Fry over medium heat for about 5 minutes, adding half the sugar during the last few minutes. Remove, rinse out the pan, and repeat the process with the remaining butter, plantains, and sugar.

Serves 6

PLANTAIN CROQUETTES (CUBA)

1 very ripe, black plantain
6 tablespoons oil
¼ pound butter
½ cup flour
2 cups milk
1 teaspoon salt
1 cup white sugar
1 teaspoon nutmeg
3 eggs, beaten
2 cups fine bread crumbs
  Oil for frying

Peel the plantain and, if desired, open it lengthwise to remove the heart. Slice crosswise in 1-inch-thick slices. Fry until golden brown in 6 tablespoons oil. Remove and drain.

Melt the butter in a saucepan. Stir in the flour and cook until the paste is golden. Gradually add the milk, which should be at room temperature. Stirring all the while to prevent sticking, cook over medium heat. When the sauce thickens, add the salt, stir, and gradually add the sugar and nutmeg. Once the sauce is thick, pour half onto a serving plate, spreading it out evenly. Place the fried plantain slices in the sauce and pour the rest of the sauce on the plantains. Let cool and refrigerate for 1 hour. Cut the dough into squares so that each holds a piece of plantain inside. Shape into croquettes. Dip twice in the egg and bread crumbs. Deep-fry at 375 degrees until the croquettes are golden brown. Remove, drain, and serve while hot.

12 to 14 croquettes

BAKED PLANTAINS

4 medium, very ripe, black plaintains

Wash and dry the plantains. Do not peel but make a lengthwise incision in the skin of each plantain, opening somewhat to show the pulp. Place on a baking sheet and cook in a preheated 350-degree oven for about 45 minutes. Turn the plantains once halfway through cooking. Serve warm in the peel.

Serves 4

PLANTAIN FRITTERS (TRINIDAD)

3 small, green plantains
2 tablespoons sugar
¾ teaspoon baking powder
   Oil for frying
1 tablespoon cinnamon

Peel each plantain and cut into 2-inch rounds. Boil them in salted water for about 30 minutes, or until tender. Remove, drain, and mash in a mortar until the pulp has a creamy consistency. Add the sugar and baking powder, mix, and shape into patties the size of a silver dollar. Fry in smoking-hot oil until golden brown. Remove, drain, and sprinkle with cinnamon.

About 12 fritters

MOFONGO (PUERTO RICO)

3 medium, ripe plantains
2 cups oil
3 cloves garlic
½ pound crisp *chicharrones*
   Salt and pepper

Peel the plantains and cut into 2-inch diagonal slices. Fry in the oil over medium heat until golden, turning once. Remove and set aside.

In a mortar, mash the garlic, adding the plantain slices one at a time while grinding. When the mixture is smooth, add the *chicharrones* and grind until well mixed. Season with salt and pepper to taste. Shape into a loaf or small balls. As a loaf, serve as a side dish with pork or ham; the balls may be served with cocktails.

Serves 4

Soups

BANANA SHRIMP BISQUE

½ cup butter
½ onion, peeled and chopped
½ cup flour
6 cups milk
2 cups chicken stock or canned chicken broth
2 cups shrimp, cooked and cleaned
1 cup cream
4 medium, ripe bananas
2 tablespoons fresh parsley, chopped
1 teaspoon nutmeg
Salt and pepper

In a saucepan, melt half the butter and fry the onion until tender. Add the flour and stir until golden. Pour in the cool milk and stir constantly until it starts to thicken. Gradually add the chicken stock. Place half the shrimp and a tablespoon or two of the cream in a blender, reduce to a thick paste, and add to the soup. Stir and simmer for 5 minutes. Peel the bananas, cut into diagonal slices, and fry to a golden color in the remaining butter. Add to the soup along with the remaining shrimp and the parsley, nutmeg, and salt and pepper to taste. Simmer another 5 minutes and pour in the rest of the cream. Simmer another minute, but *don't let the soup boil.* Serve in small bowls.

Serves 8

BANANA LENTIL SOUP

½ pound boiled ham, diced
¼ cup olive oil
2 medium onions, finely chopped
4 celery stalks, finely chopped
4 cloves garlic, minced
½ cup tomato paste
6 cups chicken stock or canned chicken broth
5 cups water
1 pound (2 cups) lentils, soaked overnight
Salt and pepper
4 green-tipped bananas

In a large, heavy-bottom pan, fry the ham in the olive oil for 2 or 3 minutes. Add the onion, celery, and garlic and cook until soft but not brown. Stir in the tomato paste and cook another 2 minutes, then pour in the chicken stock and water. Mix well and add the lentils. (If the lentils were not soaked overnight, allow an extra half-hour for cooking. In either case, always rinse the lentils before using so as to wash out pebbles and dust.) Bring the soup to a boil, cover, and simmer about 1 hour. Add salt and pepper to taste. Ten minutes before serving, peel the bananas, slice into 2-inch rounds, and drop into the soup. Serve steaming hot.

Serves 8

FRUIT SOUP (MEXICO)

1 quart very ripe strawberries
2 cups water
1 medium pineapple, peeled and cut into chunks, or 1 20-ounce can unsweetened pineapple chunks, drained
2 cups cream
1/2 cup sherry
1/2 teaspoon salt
2 teaspoons sugar
1/2 teaspoon mace
2 large, ripe bananas
  Whipped cream (optional)

Wash the strawberries and place in a saucepan with water. Bring to a boil and simmer for about 20 minutes. Remove from heat. In a blender, puree the pineapple. Remove the strawberries from the water and puree with the pineapple, adding a spoonful or two of the cooking liquid if needed. Add the puree to the saucepan with the liquid. Simmer, stirring constantly, for about 5 minutes. Pour in the cream and sherry; season with salt, sugar, and mace. Simmer for another 5 minutes, remove from heat, strain, and chill. Before serving, add the banana rounds to the soup. Garnish each serving with a dollop of whipped cream, if desired.

Serves 6 to 8

BANANA BROTH

8 large, ripe bananas
1 cup fresh orange juice
4 cups water
1 tablespoon butter
2 tablespoons sugar
1/4 teaspoon cinnamon
1/2 teaspoon vanilla extract
1/2 teaspoon salt
1/2 teaspoon grated orange rind (optional)
2 egg whites

Peel the bananas and mash with a little of the orange juice. Place the puree in a saucepan and add the remaining ingredients, except the egg whites. Stir, bring to a boil, and simmer for about 20 minutes. Remove from heat and chill well. Before serving, beat the egg whites until they form stiff peaks. Fold into the soup and serve.

Serves 6 to 8

KALLALOO (WEST INDIES)

½ pound okra, sliced
5 cups boiling water
 Juice of 2 limes
2 bunches spinach, chopped
1 ham bone, about 1 pound
¼ cup butter
2 onions, chopped
½ teaspoon thyme
½ teaspoon oregano
6 cloves garlic, minced
1 bay leaf
2 boiled crabs, cracked and
 cleaned
3 green plantains or 5 green
 bananas

Slice the okra and place in a bowl with the boiling water and lime juice. (Fresh okra should snap when broken in two; otherwise it will be too limp after cooking. Frozen okra should be defrosted.)

Place the chopped spinach and ham bone in the bottom of a large pot. In a frying pan, melt the butter and sauté the onion, thyme, oregano, garlic, and bay leaf until the onion is tender. Add to the pot, along with the crabs. Remove the okra from the lime water and add it to the pot. Bring the lime water to a rolling boil and pour it on the spinach to preserve the color. Bring to a boil again, cover, and simmer for about 1½ hours or until the ham falls off the bone. Strip the meat from the bone, return to the soup, and cook another 30 minutes.

Meanwhile, wash the plantains or bananas, cut them into 3-inch-long pieces, and boil in salted water for about 30 minutes or until tender. Remove, peel, and mash in a mortar, dipping the pestle frequently in water to prevent sticking. When the paste is smooth, shape into small balls and drop into the soup. Simmer another 5 minutes and serve.

Serves 6

EGG PLANTAIN SOUP

4 cloves garlic, minced
2 tablespoons oil
4 cups chicken stock or canned
  chicken broth
4 green plantains
  Oil for frying
4 eggs
  Salt and pepper

Fry the garlic quickly in 2 tablespoons oil in a heavy kettle. When the garlic is golden brown, add the stock, bring to a boil, and simmer, covered, for 10 minutes.

Peel and slice the plantains in half crosswise. Fry in oil for about 10 minutes at 350 degres. Remove from oil, drain, and mash in a mortar to a smooth paste. When cool enough to handle, shape into small balls (about 12). Drop the plantain balls into the simmering soup. One at a time, break the eggs onto a plate and slide each into the soup carefully so as not to break the yolk. Add salt and pepper to taste, poach for about 3 to 4 minutes, and serve immediately with slices of toast.

Serves 4

HAVANA PLANTAIN SOUP

2 green plantains
  Oil for frying
6 cups beef stock or canned
  beef broth
12 slices French bread, toasted
1½ cups grated Gruyère or Swiss
  cheese
  Salt and pepper

Peel and slice the plantains ¼ inch thick. Place the slices in salted water for 15 minutes, then drain and dry them on a kitchen towel. Deep-fry the slices in oil for 10 to 15 minutes over medium-high heat until they turn a dark gold color. Do not brown or burn them—the flavor will be spoiled. Drain the slices and, in a food processor or blender at high speed, grind them to a fine powder. (An alternative, traditional method is to grind the slices in a mortar.)

Bring the beef stock to a boil, add the plantain powder, and simmer for 10 minutes. Line the bottom of an ovenproof tureen or casserole with the toasted bread slices and sprinkle with cheese. Pour simmering soup over the bread. Place in a preheated 400-degree oven for 5 to 10 minutes, until the cheese has formed a bubbly crust. Serve immediately.

Serves 6

FISHERMAN'S SOUP (TOBAGO)

2 pounds soup bones
10 cups water
½ pound pickled pigs' knuckles
1 pound *yucas*, peeled and cut in 4-inch pieces
2 large onions, chopped
2 teaspoons thyme
1 green pepper, seeded and chopped
¼ cup butter
1½ cups raw shrimp, cleaned and deveined
3 green plantains or 5 green bananas
2 tomatoes, peeled, seeded, and chopped
2 tablespoons Tabasco sauce
Salt and pepper

Wash the soup bones and simmer in water in a large, heavy pot for about 2 hours, skimming as needed. Cut up the pigs' knuckles and add to the stock along with the *yucas*. Bring to a boil, cover, and simmer for 45 minutes or until the *yucas* are soft. Remove the *yucas*, peel away their center filaments, and mash in a mortar to a paste.

In a large frying pan, sauté the onion, thyme, and green pepper in butter until the onion is soft. Add the shrimp and fry briefly. Add the tomatoes and Tabasco sauce. Simmer for 5 minutes and pour into the soup, along with the *yuca* paste. Simmer about 30 minutes.

Meanwhile, wash and peel the plantains or bananas, cut them into 3-inch pieces, and boil them in salted water. When tender (about 30 minutes), remove, drain, and mash in a mortar, dipping the pestle in cold water frequently to prevent sticking. When the plantain paste is smooth and cool enough to handle, shape into little balls and add to the soup. Cook another 5 minutes, season with salt and pepper, and serve.

Serves 8

Meats

BANANA KABOBS

2 green-tipped bananas
Juice of 1 lemon
½ pound chicken livers, cut in
  2-inch chunks
¼ pound bacon
1 green pepper, cut in large
  pieces
Cherry tomatoes
Pearl onions
¼ cup melted butter

Peel the bananas, cut them into 3-inch chunks, and sprinkle with lemon juice. On skewers, alternate banana chunks with pieces of liver wrapped in bacon, pieces of pepper, tomatoes, and onions. Grill or barbecue, brushing occasionally with butter, for about 5 minutes, or until the liver is cooked. Serve over saffron rice.

Serves 4

BANANA VEAL MARSALA

1 pound veal scallops, thinly
  sliced
Salt and pepper
¼ cup flour
2 tablespoons butter
1 tablespoon olive oil
¾ cup Marsala
2 ripe bananas

Veal scallops should be paper thin for this dish. Ask your butcher to slice them as thinly as possible and then beat with the flat side of a mallet.

Sprinkle the veal with salt and pepper and coat with flour. In a frying pan, melt the butter and combine with the oil. Add the slices of veal and brown quickly on both sides. Pour in the Marsala and stir. Peel the bananas, slice, and add to the pan. Heat for about 5 minutes, uncovered. Serve immediately.

Serves 4

BANANA HAM ROLLS

¼ cup Dijon or French-style
  mustard
8 thin slices cured ham
8 ripe bananas
2 cups milk
2 bay leaves
  Peppercorns
¼ cup butter
¼ cup flour
6 ounces grated Swiss or mild
  cheese
  Salt and pepper
½ tablespoon fresh parsley,
  chopped

Spread mustard evenly on the slices of ham and wrap each slice around a whole peeled banana. Hold the ham in place with toothpicks. Place rolls in a buttered baking dish.

Heat the milk with the bay leaves and a few peppercorns and keep at a simmer. Melt the butter in a saucepan, add the flour, and cook until the paste is golden. Cool by plunging the saucepan into another full of cold water. Strain the simmering milk and pour all at once into the flour mixture, stirring constantly. Return to heat and simmer for about 5 minutes. When the sauce begins to thicken, add the grated cheese gradually. Season with salt and pepper to taste.

Pour the sauce over the rolls and bake in a preheated 350-degree oven for about 40 minutes. Garnish with chopped parsley.

Serves 4

GYPSY ARM

2 cups flour
¼ cup sugar
1 teaspoon salt
1 tablespoon baking powder
½ cup milk
⅓ cup butter, melted and cooled
½ recipe Picadillo Gonzalez,
  without plantains (see page
  107)
2 medium, ripe bananas
2 ripe pears, peeled and halved
¼ cup seedless raisins
1 egg white, slightly beaten

Sift the flour, sugar, salt, and baking powder together into a mixing bowl. Combine the milk with the melted butter and add all at once to the flour. Mix well, knead, and roll out on a floured board to make a rectangle approximately 10 by 12 inches.

Spread *picadillo* over the dough, in a 7-by-10-inch rectangle. Peel the bananas and place end to end down the center, with the pear halves alongside. Sprinkle with raisins. Fold the dough into a roll and pinch the edges to seal. Paint the top of the roll with egg white and place on a greased baking sheet. Bake in a preheated 425-degree oven for 25 minutes. Serve warm in slices.

Serves 6 to 8

PICADILLO ACAPULCO

1 medium onion, chopped
2 tablespoons olive oil
½ pound ground pork
½ pound ground beef
2 medium tomatoes, peeled, seeded, and chopped
1 2-inch stick cinnamon
2 pickled *jalapeño* peppers, finely chopped
¼ cup water
2 ripe bananas
2 ripe peaches, peeled and cut into chunks
1 cup pineapple chunks, fresh or canned (drained)
1 teaspoon *cilantro*
1 teaspoon cloves

Picadillo, or beef hash, is popular throughout Latin America because it transforms the often tough and dry local meat into a palatable dish. The natives of the lush and tropical state of Guerrero, Mexico, like to combine *picadillo* with the local produce—bananas, peaches, and pineapples. This recipe is one I tried in Guerrero's famous resort city, Acapulco.

In a large skillet, fry the onion in the olive oil until soft but not brown. Add the ground meat and cook until well done. Add the chopped tomatoes, cinnamon stick, *jalapeño*, and water. Cook over medium heat for about 10 minutes. Peel the bananas, slice into rounds, and add to the *picadillo*, along with the peaches and pineapple, *cilantro*, and cloves. Stir and simmer another 10 minutes. Serve over boiled rice.

Serves 4

BANANA MEAT LOAF

1 medium onion
2 cloves garlic
2 stalks celery
1½ cups unseasoned bread crumbs
¼ cup milk
1 egg
1½ pounds ground beef
3 ripe bananas
¾ teaspoon dry mustard
1 teaspoon Worcestershire sauce
2 teaspoons salt
¼ teaspoon pepper

Dice onion, garlic, and celery very fine while the bread crumbs soak in milk. Beat the egg slightly and add to the soaked bread crumbs. Combine the bread crumbs and chopped vegetables with the ground meat. Peel the bananas, mash, and mix with the mustard. Add to the meat mixture, along with the Worcestershire sauce, salt, and pepper. Shape into a loaf and place in a greased 9-by-5-by-3-inch pan. Bake in a preheated 350-degree oven for 1 to 1¼ hours.

Note: This dish is excellent for dieters, since the entire loaf contains fewer than 1,500 calories.

Serves 6

PUCHERO (MEXICO)
POT STEW

1 cup *garbanzos* (chickpeas)
2 pounds stewing beef, cut in 2-inch cubes
1 3-pound chicken, cut in serving pieces
1 pound *chorizos* or pork sausage
1 marrow bone
2 large onions, sliced
1 bay leaf
1 teaspoon thyme
2 teaspoons cloves
6 green peppers, whole
8 cloves garlic, whole
4 large zucchini, sliced
4 medium carrots, scraped and cut in large pieces
1 cup corn kernels
4 potatoes, peeled and quartered
Salt and pepper
6 large, ripe bananas
2 tablespoons butter
2 tablespoons fresh parsley, chopped

If fresh chickpeas are used, soak them overnight and drain. Place the chickpeas, along with the beef, chicken, sausage, marrow bone, onion, bay leaf, thyme, cloves, peppers, and garlic, in a large, heavy casserole or Dutch oven. Add just enough water to cover ingredients, bring to a boil, cover, and simmer for about 1 hour. Uncover the pot and add the zucchini, carrots, corn, and potatoes. Pour in more water, if needed, to cover. Bring to a boil and simmer for about 45 minutes or until the vegetables and meats are tender. Season with salt and pepper to taste.

Strain and serve the soup as a first course and the meats and vegetables as a main course. Peel the bananas, cut lengthwise into slices, and fry in butter until golden. Place the meats and vegetables in the center of a warm platter, with the slices of banana radiating like spokes. Sprinkle with chopped parsley before serving.

Serves 6 to 8

YUNGA STEW   (BOLIVIA)

3 tablespoons olive oil
4 pounds pork, cut in 1-inch cubes
4 large onions, chopped
4 cloves garlic, whole
2 green peppers, seeded and chopped
½ teaspoon cumin
5 large tomatoes, peeled, seeded, and chopped
3 cups beef stock or canned beef broth
2 fresh chili peppers (optional)
½ cup seedless raisins
2 tablespoons capers
¼ teaspoon ground saffron
¾ cup rice
5 potatoes, peeled and quartered
½ cup ground nuts (peanuts, pecans, or walnuts)
⅔ cup cream
2 tablespoons honey
4 green-tipped bananas
Salt

Bolivia, one of the two landlocked countries of South America, is divided into two regions. One is the traditional Inca center, the *altiplano*, or highlands. The other is the hot, humid region at the base of the mountain range that creates the highlands. This tropical zone, which extends into the Amazon River basin, is the *yunga*. Natives of the *altiplano*, accustomed to cold weather and rarefied air, fall ill so quickly here that the Inca emperors used the *yunga* as a pre-Columbian Siberia—a place of exile. Because of the climate, bananas have become a staple of the diet in the *yunga*.

In a heavy casserole, heat the oil and brown the pork. Remove the meat and set aside 1 tablespoon of the oil. In the remaining oil, sauté the onion, garlic, green pepper, and cumin until the onion is soft but not brown. Add the chopped tomatoes, stir, and cook 2 to 3 minutes. Return the pork to the casserole and stir in 2 cups of the beef stock. Add the chili peppers, if used, and the raisins, capers, and half the saffron. Cover and simmer for about 45 minutes.

Meanwhile, with the reserved oil, fry the rice in a pot for about 5 minutes, without burning. Stir with a fork so the grains won't break. Then add 1 cup beef stock and the remaining saffron, stir, and bring to a boil. Cover tightly and simmer for about 20 minutes, or until the rice is fluffy but not overcooked. Remove from heat.

When the stew has cooked about 45 minutes, add the potatoes and simmer 20 to 30 more minutes. Mix in the rice, ground nuts, cream, and honey. Stir. Peel the bananas, cut into rounds, and add to the stew. Cover and simmer 10 to 15 more minutes. Season with salt to taste.

Serves 8

BANANA RICE HASH

1 large onion, chopped
1 green pepper, chopped
⅓ cup peanut oil
2 cups rice
1 pound ground beef
3 cups chicken stock or canned
   chicken broth
¼ teaspoon oregano
¼ teaspoon cumin
   Salt and pepper
4 ripe bananas
2 tablespoons butter
4 eggs

In a frying pan, sauté the onion and green pepper in oil until soft. Add the rice and stir to coat well. Add the ground beef and cook for 4 to 5 minutes. Pour in the chicken stock and seasonings and stir with a fork. Cover tightly and simmer for 15 minutes. Meanwhile, peel the bananas, slice lengthwise, and sauté in butter. Beat the eggs well, uncover the hash, and mix in well. Cook, uncovered, for another 5 minutes, stirring constantly. Place the hash on a platter and surround with fried bananas.

Serves 6

GOVERNOR'S PIE   (CUBA)

2 recipes Pastry I or II, unbaked
   (see pages 159–60)
2 teaspoons Dijon or French-
   style mustard
½ pound cured ham, sliced
½ pound roast pork, sliced
¼ pound turkey breast, sliced
2 fairly green bananas, sliced
   lengthwise
½ pound Muenster cheese, sliced
1 dill pickle, sliced very thin
1 egg yolk
1 tablespoon milk

Use one recipe of pastry to line the bottom of a greased 8-inch pie pan. Coat the bottom with mustard. Make successive layers of ham, pork, turkey, bananas, cheese, and pickle. Cover with the other half of the dough. Crimp the edges with the tines of a fork and make some small slits in the top. Brush the dough with a mixture of egg yolk and milk. Bake in a preheated 425-degree oven for 20 to 30 minutes or until the crust is golden.

Serves 6

GROUNDNUT STEW (WEST AFRICA)

2 pounds stewing beef, cut in
    1-inch cubes
4 onions, sliced
7 cups water
1 3-pound chicken, cut in serv-
    ing pieces
1 teaspoon thyme
1 teaspoon sage
1 cup peanut butter
4 large tomatoes, peeled,
    seeded, and chopped
2 teaspoons crushed red pepper
6 hard-boiled eggs, chopped
¼ cup fresh parsley, chopped
6 cloves garlic
1 green pepper, seeded and
    chopped
    Salt
6 green-tipped bananas
¼ cup oil

British colonists called peanuts groundnuts since, naturally enough, the plant is a sort of creeper. In much of Africa peanuts are used to thicken stews, as is the case in this dish.

In a large, heavy casserole, place the beef, 2 onions, and 2 cups of water. Bring to a boil, then simmer for about 30 minutes or until the water begins to evaporate and the meat starts to brown. Add the chicken and another cup of water. Again wait until the chicken is browned and then pour in the remaining water, along with the thyme and sage. Bring to a boil, cover, and simmer for 15 minutes. Using some of the cooking liquid, blend the peanut butter into a smooth paste and add gradually to the meat, stirring all the while. Add the tomatoes, red pepper, eggs, parsley, and remaining onion, garlic, and green pepper. Season with salt to taste. Boil for about 10 minutes, then cover and simmer 45 minutes or until the beef is tender.

Meanwhile, peel the bananas, cut in thick diagonal slices, and brown in oil. Remove and drain on paper towels. Add to the stew 5 minutes before serving. Serve over plain boiled rice.

Serves 10

PICADILLO GONZALEZ (CUBA)

¼ cup olive oil
1 large onion, chopped
1 large green pepper, seeded and chopped
2 cloves garlic, chopped
1 pound ground sirloin
1 pound ground veal
1 large tomato, peeled, seeded, and chopped
½ teaspoon ground cumin
½ cup tomato sauce
½ tablespoon capers
¼ cup pimiento-stuffed olives, sliced
½ cup seedless raisins
Salt and pepper
2 ripe plantains
Oil for frying
12 eggs

Heat the olive oil in a frying pan and sauté the onion, green pepper, and garlic over medium heat until the onion is soft. Add the ground sirloin and veal and cook for 4 to 5 minutes. Add the tomato, cumin, tomato sauce, capers, olives, raisins, and salt and pepper to taste. Mix thoroughly. Cover and simmer for 20 minutes.

Meanwhile, peel the plantains and slice in 1-inch-thick diagonal pieces. Fry in oil over medium heat until golden brown.

Serve the *picadillo* on individual plates: Spoon *picadillo* over a bed of boiled white rice. Arrange the plantain slices like spokes over the *picadillo*, and top each serving with 2 fried eggs.

Serves 6

PIOUS NUNS (PUERTO RICO)

3 ripe plantains
3 tablespoons oil
1 medium onion, finely chopped
2 cloves garlic, minced
1 pound ground sirloin
2 medium tomatoes, peeled, seeded, and chopped
⅛ teaspoon saffron
½ teaspoon oregano
1 tablespoon capers
8 pimiento-stuffed olives, sliced
1 teaspoon salt
3 eggs, beaten
1 cup bread crumbs
   Oil for frying

Peel and cut each plantain into 4 lengthwise slices. Fry in oil until golden brown. Remove, drain on paper towels, and carefully form a circle with each slice, holding it in place with toothpicks. In a separate frying pan, sauté the onion and garlic in oil until the onion is soft. Add the ground sirloin and cook until it loses its red color. Add the tomatoes, saffron, oregano, capers, olives, and salt. Stir, cover, and simmer for about 10 minutes. Remove from heat.

When the ground meat is cool, spoon it into the plantain circles. Spread beaten egg over the top and sides of each circle and sprinkle with bread crumbs. Place a small dish carefully over each circle and flip over. Spread beaten egg and sprinkle bread crumbs on the other side. When dry, coat each plantain circle with egg and bread crumbs again. Deep-fry each stuffed plantain in hot oil until brown. Serve hot.

Note: Instead of egg and bread crumbs, the following paste may be used:

¾ cup flour
1 teaspoon baking powder
¼ teaspoon salt
1 egg
⅓ cup milk

Serves 4

WEST INDIAN BEEF STEW

2 pounds boneless stewing beef,
cut in 2-inch cubes
2 teaspoons ground black
pepper
2 teaspoons thyme
1 teaspoon paprika
⅓ cup flour
1 tablespoon brown sugar
½ cup oil
2 large onions, sliced
2 large tomatoes, peeled,
seeded, and chopped
4 cups water
3 tablespoons vinegar
2 large, green plantains or 4
green bananas

Sprinkle the beef cubes with pepper, thyme, and paprika. Mix well and coat with flour. In a heavy casserole, heat the oil with the sugar over medium heat until the sugar melts and bubbles. *Do not let the sugar burn.* Brown the beef in the sugar oil and remove. In the same oil, sauté the onions until tender and then add the chopped tomatoes. Cook for about 1 minute. Return the beef to the casserole and add the water and vinegar. Bring to a boil, cover, and simmer for 1½ hours.

Meanwhile, wash the plantains and slice into 3-inch-thick pieces. Boil in salted water for 20 minutes or until tender. Drain, peel, and mash in a mortar. Dip the pestle frequently in cold water to avoid sticking. When the plantains are mashed and cool enough to handle, shape into balls and drop into the stew. Cook 15 minutes more. Serve the stew in bowls with plain boiled rice.

Serves 6

ADJUNTAS BAKED PIE (PUERTO RICO)

4 medium, ripe plantains
1 cup milk
Salt and pepper
½ cup butter
1 large onion, chopped
4 cloves garlic, minced
1 green pepper, seeded and
chopped
1 pound ground sirloin
3 tablespoons seedless raisins
1 tablespoon capers
¼ cup pimiento-stuffed olives
3 hard-boiled eggs, sliced

Wash and peel the plantains. Cut in 2-inch-thick slices and boil in salted water 20 to 30 minutes or until tender. Drain and mash. Heat the milk and stir in the mashed plantains to make a soft but thick paste. Add salt and pepper to taste.

In a frying pan, melt the butter and sauté the onion, garlic, and green pepper until tender. Add the ground sirloin and cook over medium heat until the meat is well done. Add the raisins, capers, olives, sliced eggs, and salt and pepper to taste. Mix well, cover, and simmer for about 10 minutes.

Spread half the plantain paste on the bottom of a well-greased baking dish. Spoon the meat mixture over the layer of plantain and cover with the remaining paste. Bake in a pre-heated 350-degree oven for 30 minutes or until the crust is crisp. Recommended for lunch with tossed salad and white wine.

Serves 4

PLANTAIN POT

1 pound boneless pork, cut in
  2-inch cubes
¼ cup oil
2 onions, chopped
4 cloves garlic
1 green pepper, seeded and
  chopped
3 large tomatoes, peeled,
  seeded, and chopped
2 teaspoons ground red pepper
1 quart water
4 ripe plantains
2 medium-sized smoked herrings
  Juice of 1 lime
  Salt

In a heavy casserole, brown the pork in the oil. Remove the pork and sauté the onions, garlic, and green pepper in the remaining oil until soft. Add the chopped tomatoes and red pepper, stir, and cook for 3 minutes. Return the pork and add the water. Bring to a boil, cover, and simmer for 1 hour.

Peel the plantains, chop into 2-inch cubes, and add to the stew. Simmer 20 minutes. Chop up the herrings, wash in the lime juice, and add to the stew. Cover and simmer 15 to 20 more minutes. If you want a thicker stew, mash up some plantains and add. Season with salt to taste. Serve over plain boiled rice.

Serves 8

MACHUQUILLO PORK CHOPS (CUBA)

4 cloves garlic
1 teaspoon salt
½ teaspoon black pepper
  Juice of 1 lime
6 pork chops
1 onion, sliced
½ cup water
¼ cup dry white wine
5 green plantains
  Oil for frying

In a mortar, pound 3 garlic cloves with the salt and pepper and the lime juice. Pour over the pork chops. Cover the meat with the sliced onion and let stand for about 1 hour. Drain the chops and place in a frying pan with the water and wine. Cook until the liquid evaporates and the pork chops are golden brown, about 30 minutes.

Meanwhile, peel the plantains, cut in 2-inch slices, and fry in hot oil until golden. Drain, place on wax paper, and flatten with the heel of your hand. Return the slices to the pan and fry until crisp. Remove, chop up, and place in a blender. Add 3 tablespoons of fat drained off the pork chops and reduce the plantains to a paste. Transfer the paste to a skillet with 1 tablespoon oil and 1 minced garlic clove, sprinkle with salt, and cook over a medium flame for a few minutes, stirring all the while to prevent sticking. Serve the pork chops over the plantain paste.

Serves 6

LEAFY LAMB (ZAIRE)

¼ cup oil
2 pounds stewing lamb, cut in 2-inch cubes
1 onion, chopped
3 cloves garlic, minced
½ cup tomato sauce or 2 tomatoes, peeled, seeded, and chopped
¼ teaspoon thyme
¼ teaspoon sage
½ teaspoon black pepper
Juice of 1 lime
2 cups water
2 ripe plantains
1 slice lime
1 bunch spinach
1 bunch watercress
1 bunch collard greens

In a large casserole, heat the oil, brown the lamb cubes, and remove. In the same oil, fry the onion and garlic until soft. Return the lamb, add the tomatoes or tomato sauce, thyme, sage, pepper, lime juice, and 1 cup water. Stir, bring to a boil, cover, and simmer for about 1 hour.

Meanwhile, wash the plantains, slice into 3-inch chunks, and boil in water with a slice of lime for 30 minutes or until tender. Remove, drain, peel, and set aside.

When the lamb is tender, remove it from the casserole and keep it warm. Clean all the greens, chop them up coarsely, and add to the cooking sauce. Pour in 1 cup water, cover, and simmer about 15 minutes. Add the plantains and cook 5 minutes more or until warmed through. When the greens are cooked and the sauce has thickened, place the lamb cubes on a bed of rice on a serving platter and pour the green sauce over.

Serves 4

MONDONGO SABROSO (PUERTO RICO)
TASTY TRIPE

4 pounds tripe
¾ cup oil
2 tablespoons *achiote* seeds
½ pound salt pork, diced
2 large onions, chopped
1 large green pepper, roasted,
    peeled, seeded, and sliced
1 head of garlic, peeled and
    minced
2 cups tomato sauce
½ cup dark rum or brandy
1 pickled hot green pepper,
    chopped
1 teaspoon oregano
1 tablespoon *cilantro*
1 tablespoon basil
1 tablespoon chopped fresh
    parsley
Juice of 1 lime
Salt and pepper
2 ripe plantains
1 pound potatoes, peeled and
    sliced
1 cup cooked chickpeas
1 tablespoon vinegar

Clean out the tripe, blanching it for a few minutes each time in three changes of boiling water. Place it in a large, heavy casserole, cover with water, and cook for 2 to 3 hours or until tender. Drain and reserve the liquid. Cut the tripe into 2-inch-square slices and set aside. Rinse out the casserole and dry.

Heat the oil in the casserole and cook the *achiote* seeds for 2 or 3 minutes, until the oil turns a reddish color. Remove the seeds. Sauté the cut-up tripe and diced salt pork and remove. In the remaining oil fry the onions, green pepper, and garlic until soft. Pour in the tomato sauce, stir, and cook 3 to 4 minutes. Add the rum or brandy. Stir, adding the hot pepper, oregano, *cilantro*, basil, parsley, lime juice, and salt to taste. Cook over medium heat for 10 minutes, stirring frequently. Return the tripe and pork. Pour in 2 cups of cooking liquid, stir, and bring to a boil.

Peel the plantains, cut into 2-inch-thick slices, and drop into the stew. Cover the casserole and simmer for 10 minutes. Add the sliced potatoes and the chickpeas. Cover and simmer until the vegetables are tender and the stew has thickened. Before serving, add the vinegar and cook 2 to 3 minutes more. Add salt and pepper to taste. If the stew is too thin, uncover the casserole and cook over medium heat for another 5 minutes.

Serves 8 to 10

CHILINDRON DE CHIVO (CUBA)
SPICY GOAT STEW

8 pounds kid or goat meat, or
 6 pounds boneless lamb
 Juice of 6 limes
6 medium onions (4 chopped,
 2 sliced)
3 heads of garlic
4 fresh green chilies (2 whole,
 2 seeded and chopped)
¼ cup *cilantro*
6 bay leaves
2 tablespoons oregano
¼ cup fresh parsley, chopped
5 quarts water
4 green peppers, seeded and
 chopped
1¼ cups olive oil
6 ounces dark rum
½ cup tomato paste
2 pounds tomatoes, peeled,
 seeded, and chopped
4 ripe plantains
8 green plantains
3 pounds potatoes, peeled and
 sliced
2 pounds *malangas*, peeled and
 sliced
1 pound *calabazas* or Hubbard
 squash, peeled and quartered
 Salt
4 tablespoons cider vinegar

This is a typical Cuban *guajiro*, or peasant dish eaten on feast days with family and friends. *Chilindron* is the Cuban version of *enchilada*, that is, a dish cooked with chilies. It is the exception, rather than the rule, of Cuban cooking. Kid or, more commonly, goat meat is used, the latter being a tougher but tastier meat. However, lamb makes a very pleasant substitute.

Marinate the meat overnight in the lime juice with 2 sliced onions, a peeled head of garlic cloves, and 1 chili. Drain before cooking. Place the meat, sliced onions, garlic, and chili in a heavy casserole large enough to hold all the ingredients. Add another whole chili and the *cilantro*, bay leaves, oregano, parsley, and water. Bring to a boil and then simmer gently for 1½ to 2 hours or until the meat is tender. Skim the surface occasionally.

Remove the meat from the casserole, strain, and reserve the liquid. Rinse out the casserole and dry. Chop up the remaining onions and sauté, along with the green peppers, 2 peeled heads of garlic and 2 chopped chilies in olive oil in the casserole until the onions are soft. Stir the rum into the tomato paste, add it and the chopped tomatoes to the casserole, and cook 3 to 4 minutes. Return the meat to the casserole and pour in 1 quart of the reserved liquid. Bring to a boil.

Meanwhile, peel the plantains and cut into 2-inch-thick slices. Drop into the soup, along with the potatoes, *malangas*, and *calabazas*. Cover the casserole and simmer for 45 minutes or until the vegetables are tender and the *calabaza* has dissolved, thickening the stew. Salt to taste. Before serving, mix in the cider vinegar. Cook 3 minutes longer. Serve over plain boiled rice in soup bowls.

Serves 20

OXTAIL STEW (PUERTO RICO)

½ cup olive oil
3 onions, chopped
4 cloves garlic, minced
4 pounds oxtail pieces
½ cup light rum
4 quarts water
1 teaspoon fresh parsley, chopped
1 tablespoon *cilantro*, chopped
1 teaspoon ground oregano
  Juice of 2 limes
1 pickled hot green pepper, chopped (optional)
2 green peppers, roasted, seeded, and sliced
1 pound *yucas*, peeled and cut in 3-inch pieces
2 pounds *yautias*, peeled and cubed
1 pound *calabazas* or Hubbard squash, peeled and cubed
6 *guineos verdes* or 3 medium plantains
  Salt and pepper

This dish is ordinarily made with a type of plantain called *guineo verde*, smaller and starchier than a regular plantain but similar in size and shape to a standard banana. Plantains, however, may be used instead.

Heat the oil in a large, heavy casserole. Sauté the onions and garlic until soft. Add the oxtail pieces and fry briefly. Pour in the rum and stir until the alcohol evaporates. Pour in the water and add the parsley, *cilantro*, oregano, lime juice, and hot pepper, if used. Bring to a boil, cover, and simmer for 2½ hours.

Uncover and add the green pepper, *yucas*, *yautias*, and *calabazas*. Peel the *guineos verdes* and cut each into 3 pieces—or cut the plantains into 2-inch slices—and add to the stew. Cover and simmer another hour, until the *calabaza* has dissolved and thickened the stew and the vegetables are tender. Add salt and pepper to taste.

Serves 6 to 8

FIELD HAND STEW (ARUBA)

1 cup dried chickpeas
4 quarts water
½ pound salt pork, diced
4 pounds stewing beef, cut in cubes
2 large onions, chopped
1 pound *yautias*, peeled and cubed
1 pound *yucas*, peeled and cubed
1 whole hot pepper
2 green plantains
2 medium-ripe plantains
1 very ripe plantain
1 cup coconut cream
Salt and pepper

Soak the chickpeas in the water overnight. In a large, heavy casserole, try out the salt pork. Sauté the beef in the rendered fat and remove. Fry the onions in the fat until tender. Pour in the chickpeas with their water and return the beef to the casserole. Bring to a boil, cover, and simmer for 3 hours or until the chickpeas are tender.

Uncover and add the *yautias*, *yucas*, and hot pepper. Peel the plantains, cut into 2-inch slices, and add to the stew, along with the coconut cream. Bring to a boil and simmer for 30 minutes or until the vegetables are tender. Season with salt and pepper to taste.

Serves 8

PASTELES (PUERTO RICO)

Dough:

4 pounds green plantains
(about 4 medium)
2 pounds *yautias**

¾ cup *achiote* lard
1 tablespoon salt
Chicken stock or canned
chicken broth (about 2 cups)

Filling:

1 pound lean boneless pork,
finely diced
3 tablespoons lime juice
½ cup *achiote* lard
1½ pounds lean boneless cured
ham, finely diced
1 medium onion, ground or
finely chopped
2 cloves garlic, minced
¼ pound salt pork
1½ cups cooked chickpeas
(*garbanzos*)
3 tablespoons capers
24 pitted green olives, sliced
3 roasted pimientos, finely
chopped
3 tomatoes, peeled and
chopped
2 green peppers, finely
chopped
¼ cup seedless raisins (optional)

½ cup slivered almonds
1 teaspoon ground oregano
1 tablespoon *cilantro*, finely
chopped
1 cup chicken stock or water
Parchment paper or banana
leaves

Pasteles are a typical Puerto Rican treat during Christmas time. The island custom on Christmas Eve is to pay a visit to the homes in the neighborhood and sing *aguinaldos*, or carols, in exchange for food and drink. After the performance, the person whose house is visited invites the carolers inside to rest from their strenuous journey—usually the distance from the house next door. The host traditionally serves *pasteles* and other treats, along with braces of rum.

Peel the plantains and *yautias* and grate coarsely. Place in an electric blender and add the *achiote* lard, salt, and enough chicken stock to make a smooth, thick paste. Set aside until the filling is prepared.

* *Yautias* are not the essential ingredient; plantains are. Therefore they may be substituted, if desired, with the same weight in green plantains.

Rub the pork with the lime juice. In a large casserole, heat the *achiote* lard and brown the pork. Add all the remaining ingredients, stir well, bring to a boil, cover, and simmer for 20 minutes. Set aside until cool enough to handle.

Pasteles are traditionally wrapped in the large leaves of the banana plant. If leaves are available, wash them well and remove the spines in the middle for greater flexibility. Cut the leaves into 12-inch squares and grease each square lightly on one side. If banana leaves are not available, squares of greased parchment paper of the same proportions may be used. Allow 2 pieces of wrapping per *pastel*, that is, about 48 pieces for this recipe. It's always handy to have a few extra in case of tearing.

Place 3 tablespoons of dough in the middle of a wrapper. Spread out the dough to make a smooth and very thin layer approximately 6 by 8 inches. Spread 3 tablespoons of the filling lengthwise down the center of the dough. Fold the wrapper lengthwise and press the edges to seal the dough. Fold the short ends toward the center. Wrap in a second wrapper placed on the diagonal.

Repeat this process for another *pastel* and tie both *pasteles* together with kitchen string, placing folded edges against each other. Continue making *pasteles* until the ingredients are used up.

Into a large pot, pour 5 quarts of water with 4 tablespoons salt for every 12 *pasteles*. Bring the water to a boil and drop the *pasteles* in. Bring to a boil again, cover, and simmer for 1 hour, turning the *pasteles* once while cooking. Remove the *pasteles* from the water, drain, cut the strings, and serve steaming hot in their wrappers. *Pasteles* may be seasoned with Tabasco or any kind of hot sauce, if desired.

24 pasteles

Poultry

DOMINICAN CHICKEN (DOMINICAN REPUBLIC)

¼ cup oil
1 medium onion, chopped
4 cloves garlic, minced
¼ cup flour
1 3- to 3½-pound chicken, quartered
¼ cup dark rum
1½ cups water
1½ cups pineapple juice
1 pickled pepper (*jalapeño* style), chopped
1 teaspoon aniseed
1 4-inch stick cinnamon
2 teaspoons basil
Salt and pepper
⅛ teaspoon saffron
3 cups pineapple chunks
3 large, green-tipped bananas, cut in 2-inch chunks
¼ cup slivered almonds

Heat the oil in a large, heavy casserole. Sauté the onion and garlic until soft and remove. Put the flour in a paper or plastic bag, place the chicken quarters in the bag, and shake until coated with flour. Dust off the excess flour and brown the chicken in the casserole. Remove the chicken, pour off the excess oil, and deglaze the bottom of the casserole with the rum. Add water and return the chicken, onions, and garlic. Pour in the pineapple juice and add the pickled pepper, aniseed, cinnamon, basil, and salt and pepper to taste. Then add the powdered saffron and mix until a medium yellow color. Bring to a boil, cover, and simmer for 1 hour or until the chicken is tender.

Uncover, add the pineapple chunks, banana, and almonds, and simmer an additional 10 to 15 minutes. Place the chicken on lettuce leaves on a serving platter, with pineapple and banana pieces as garnish. Serve with boiled rice.

Serves 4

TOBAGO PELAU (WEST INDIES)

1 2½- to 3-pound stewing
  chicken, cut in serving pieces
2 teaspoons thyme
2 teaspoons paprika
2 teaspoons ground black
  pepper
1 cup flour
¾ cup oil
1½ tablespoons brown sugar
1 large onion, chopped
1 green pepper, seeded and
  chopped
2 large tomatoes, peeled,
  seeded, and chopped
5 cups water or half water, half
  chicken broth
2 cups rice, washed
¼ teaspoon ground saffron
5 green-tipped bananas
¼ cup butter
1 teaspoon ground cinnamon
¼ cup slivered almonds
  Salt
¼ cup sliced pimiento-stuffed
  olives

Sprinkle the chicken with thyme, paprika, and pepper and coat with flour. In a large, heavy casserole, heat the oil with the sugar over medium heat until a bubbling foam forms on top. *Do not let the sugar burn*, or it will impart a bitter taste. Brown the chicken in the sugar oil and remove.

In the same oil, fry the onions and green pepper until soft but not brown. Then add the chopped tomatoes, simmer about 1 minute, and pour in the water or water and chicken stock. Bring to a boil, cover, and simmer for 45 minutes or until the chicken is tender. Add the rice and saffron, stir well with a kitchen fork so the grains won't break, and bring to a boil. Cover tightly and simmer for 20 minutes.

Meanwhile, peel the bananas and slice each diagonally into 5 pieces. Sauté in half the butter and sprinkle with cinnamon as they cook. Remove and keep warm.

When the rice is done, the grains should be fluffy but not mushy. Add the remaining butter and almonds. Add salt to taste. Then mix in. Place the chicken with rice on a serving platter and garnish with olives and the cinnamon bananas.

Serves 8

TOM TOM CHICKEN (HAITI)

3 bananas
    Juice and grated rind of 1 lime
1 cup bread crumbs
1 teaspoon brown sugar
¼ teaspoon nutmeg
½ teaspoon cayenne
½ teaspoon thyme
7 tablespoons dark rum
    Salt and pepper
1 4-pound chicken
¼ cup butter
½ cup chicken broth

Peel and slice the bananas, then sprinkle with lime juice. Combine the bread crumbs with the bananas, sugar, nutmeg, grated lime rind, cayenne, thyme, 3 tablespoons rum, and salt and pepper to taste. Stuff the chicken with this mixture, truss, and place in a roasting pan. Roast in a preheated 375-degree oven for about 1½ hours, basting frequently with butter.

When the chicken is done, remove it from the oven, cut the trussing strings, and place on a warm serving platter. Add the chicken broth to the pan juices, along with 2 tablespoons rum. Cook over high heat, stirring to remove attached food bits, until the gravy is thickened somewhat. Serve in sauce boat. At the moment of serving, heat up the remaining rum, pour over the chicken, and flame.

Serves 4

CHICKEN GUMBO (CUBA)

1 pound okra, fresh or frozen
3 cups water
    Juice of 3 limes
⅓ cup oil
1 2½-pound chicken, cut in
    serving pieces
1 onion, chopped
1 green pepper, chopped
2 cloves garlic, minced
1 cup tomato sauce
⅓ cup vinegar
1 cup chicken stock
2 cups dry white wine
2 teaspoons salt
½ teaspoon pepper
¼ cup dried shrimp (optional)
3 medium-ripe plantains

Okra, if bought fresh, should snap when broken; otherwise, it will get too soggy when cooking. Frozen okra should be defrosted. In either case, the okra should be sliced and placed in a bowl with the water and lime juice.

In a large casserole or Dutch oven, heat the oil, brown the chicken, and remove. Sauté the onion, green pepper, and garlic in the remaining oil until tender. Add the tomato sauce, vinegar, chicken stock, wine, salt, and pepper. Return the chicken to the pot and simmer for 30 minutes or until the chicken is half cooked. Add the dried shrimp, if desired.

Meanwhile, cut the plantains into 2-inch-thick slices and boil in salted water for 20 minutes or until tender. Remove, peel, and mash. Once the plantains have cooled, shape into small balls and set aside in a dish.

Add the okra with lime water to the stew. Simmer for 20 to 30 minutes or until the okra is tender and cooked. Add the plantain balls, cover, and simmer 5 more minutes. Serve in soup bowls, with 2 tablespoons boiled rice on top, if desired.

Note: For a thicker gumbo, drain the lime water from the okra and simply add 1 ½ cups plain water.

Serves 6

GALLINA EN JARDIN (CARIBBEAN)
CHICKEN IN GARDEN

2 3-pound chickens, cut in
  serving pieces
2 large onions, each stuck with 2
  cloves
2 large tomatoes, peeled
1 carrot, scraped
1 medium-sized ham bone
1 fresh chili, whole
1 garlic head
4 quarts water
2 teaspoons leaf oregano
1 tablespoon fresh parsley,
  chopped
1 pound potatoes, peeled and
  sliced
1 pound *malangas*, peeled and
  cubed
4 ears sweet corn, each cut in 3
  pieces
1 pound *calabazas* or Hubbard
  squash, peeled and cubed
4 ripe plantains
1 medium head cabbage, cut
  in wedges
1 cup sweet peas
¼ cup sliced green olives
¼ teaspoon saffron

2 tablespoons cider vinegar
  Salt and pepper

I have no idea how this recipe got its name; perhaps the garden refers to the vegetables in the stew. *Gallina* refers to a fully grown hen, which is ordinarily used in this recipe. I have cut down the cooking time to adjust it to American chickens.

Place the pieces of chicken in a heavy casserole. Add the onions, tomatoes, carrot, ham bone, chili, unpeeled garlic head, and water. Bring to boil, cover, and simmer for about 1 hour, skimming occasionally. Remove the chicken, strain the broth, and reserve the liquid. Scrape bits of ham from the bone and reserve. Discard the bone.

Rinse out the casserole and return the chicken, ham bits, and broth to it. Add the oregano, parsley, potatoes, *malangas*, corn, and *calabazas*. Bring to a boil, cover, and simmer for 20 minutes. Peel the plantains, slice into 3-inch-thick chunks, and add to the stew. Bring to a boil, cover again, and simmer for 15 minutes.

Uncover, add the cabbage, peas, olives, and saffron. Cover, and simmer again for 10 minutes more. The vegetables should be tender and the *calabazas* should have distintegrated and thickened the stew. Two minutes before serving, add the vinegar and salt and pepper to taste, and stir.

Serves 12

Fish
and Eggs

BANANA BLACK-EYED PEAS

2 cups (1 pound) dried black-
eyed peas
2 large onions, chopped
1 green pepper, seeded and
sliced
2 cloves garlic, minced
¼ cup oil
¼ cup tomato paste
2 teaspoons crushed red pepper
2 large tomatoes, peeled and
seeded
½ pound shrimp, shelled and deveined
Salt
4 large, green-tipped bananas
Oil for frying

Wash the peas and soak in water overnight. The next day, boil them in 2 quarts water until tender, about 1½ hours. Drain the peas and set aside.

In a large frying pan, sauté the onions, green pepper, and garlic in the oil until tender. Add the tomato paste, red pepper, tomatoes, and shrimp. Simmer for 15 minutes. Add the peas. Bring to a boil, cover, and simmer 15 minutes without stirring. Uncover and simmer 5 minutes, until thick. Add salt to taste.

Meanwhile, peel the bananas, cut into 2-inch-thick diagonal slices, and deep-fry in oil until golden brown. Serve with the peas.

Serves 6

BANANA-STUFFED BASS

¼ cup chopped onion
¼ cup butter
1 cup unseasoned bread cubes
Juice and grated rind of 1 lime
½ teaspoon salt
½ teaspoon pepper
1 teaspoon grated fresh ginger
2 green-tipped bananas
1 3- to 4-pound sea bass, split
and boned

Sauce:
2 tablespoons butter
¼ pound mushrooms, chopped
1 cup tomato sauce
¼ cup tomato paste
1 tablespoon lime juice
½ teaspoon pepper
¼ cup fresh parsley, chopped
Slices of lime

Sauté the onion in melted butter until soft. Add the bread cubes, lime rind, salt, pepper, and ginger and mix. Peel the bananas, slice lengthwise, and sprinkle with lime juice.

Wash and dry the fish. Stuff the banana slices in the bass, placing them against the backbone, then stuff the bread cube mixture around the bananas. Place the bass in a greased baking dish.

In a saucepan, melt 2 tablespoons butter and sauté the mushrooms briefly. Add the tomato sauce, tomato paste, lime juice, and pepper to taste. Mix well, bring to a boil, and pour over the fish. Sprinkle with parsley.

Bake the bass in a preheated 375-degree oven for 30 to 40 minutes or until the flesh flakes with a fork. Garnish with lime slices.

Serves 6

BANANA SOLE (PHILIPPINES)

3 green-tipped bananas
Juice and grated rind of 1 lime
½ cup oil
1 medium onion, chopped
2 cloves garlic
1 green pepper, sliced
¼ teaspoon cayenne
½ teaspoon grated fresh ginger
½ teaspoon salt
2 tablespoons sugar
1 cup water
2 pounds sole fillets
Salt and pepper
¼ cup flour
1 tablespoon cornstarch
¼ cup fresh parsley

Peel the bananas and cut into thick slices. Sprinkle with lime juice and set aside. In 2 tablespoons oil, sauté the onion, garlic, and green pepper slices until soft. Add the lime rind, cayenne, ginger, salt, and sugar and cook over medium heat for about 2 minutes. Add water to the sauce, bring to a boil, cover, and simmer.

Dust the sole fillets with salt and pepper and coat with flour. In the remaining oil, fry the sole. When the flesh flakes easily, remove to a serving platter and keep warm.

Combine the cornstarch with 1 tablespoon water and add to the sauce to thicken. Also add the banana slices and parsley, then stir and simmer for 5 minutes. Ladle the sauce over the sole and serve.

Serves 4

BANANA OMELETTE (SOUTH AMERICA)

6 ripe bananas
⅓ cup butter
6 eggs, separated
2 teaspoons salt
⅓ cup cream
¼ teaspoon cayenne
1 cup grated Muenster cheese
1 tablespoon fresh parsley, chopped
Black pepper

Peel the bananas and slice in half lengthwise and then crosswise. In a frying pan, melt the butter and fry the bananas on both sides until golden brown. Remove and drain.

Beat the egg yolks with salt, cream, and cayenne. When well beaten, add the cheese and mix well. Beat the egg whites separately until they form moist, stiff peaks. Fold into the yolk mixture.

Grease a baking dish well and pour in the eggs. Arrange the bananas on top. Bake in a preheated 350-degree oven for 20 minutes. Remove and garnish with parsley and black pepper to taste.

Serves 4

MAYAGUEZ ROSE (PUERTO RICO)

4 green plantains
2 tablespoons lemon juice
2 tablespoons olive oil
¼ cup milk
Salt and pepper
3 hard-boiled eggs
2 cups sliced green peppers
3 sweet red peppers, sliced in long strips
½ cup pimiento-stuffed olives
1 whole radish

Wash the plantains, cut each into 3 pieces, and boil in water with lemon juice for 20 to 30 minutes or until tender. Peel and mash the plantains, mixing with the oil and milk to a creamy paste. Add salt and pepper to taste. Spread the plantain mixture on a large serving plate.

Slice the hard-boiled eggs and separate the yolks and whites. Chop the yolks finely and sprinkle in the center of the plantain. Arrange the sliced whites in a ring around the chopped yolks. Next, place green pepper slices in a ring around the egg whites and then red pepper, so the plate resembles a rose. Olives form the outermost ring. Garnish with a radish in the center of the rose. Serve cold.

Serves 6

PIÑÓN (PUERTO RICO)
BEEF PLANTAIN OMELETTE

3 very ripe plantains
  Oil for frying
1 onion, chopped
½ green pepper, chopped
2 cloves garlic
½ pound ground beef
¼ cup tomato sauce
1 tablespoon capers
1 tablespoon sliced green olives
  (optional)
  Salt and pepper
½ pound string beans, fresh or
  frozen, cut in 3-inch pieces
6 eggs
¼ cup butter

This is my favorite plantain recipe. Although a little difficult, it's worth the effort.

Peel the plantains, cut into 2-inch-thick lengthwise slices, and fry in oil until golden brown. Remove, drain, and keep warm.

In a frying pan, sauté the onion, green pepper, and garlic until soft but not brown. Add the ground beef and fry at high heat for 3 minutes. Pour in the tomato sauce and add the capers and olives, if desired. Cook 15 minutes over medium heat, stirring occasionally. Season with salt and pepper to taste.

Wash the string beans and boil in salted water for 20 minutes or until tender. Remove and drain well.

Beat the eggs, adding salt and pepper to taste. Butter the sides and bottom of a round casserole and melt the remaining butter in the bottom.

Pour in *half* of the beaten eggs and cook over medium heat for about 1 minute or until slightly set. Cover the eggs with one-third of the plantain slices, following with layers of half the ground meat and half the string beans. Add another layer of plantains, the remainder of the ground beef, another layer of beans, and top with plantains. Pour the rest of the beaten eggs over the top.

Cook over low heat for 15 minutes, uncovered, being careful not to let the omelette burn. Then place in a preheated 350-degree oven for 10 to 15 minutes to brown the top of the *piñón*. Serve with rice and beans. Excellent for lunch.

Serves 4

CODFISH SERENADE (PUERTO RICO)

2 pounds salted codfish fillets
2 large Bermuda onions
3 large tomatoes, sliced
2 green peppers, roasted,
  seeded, and sliced
4 cloves garlic, peeled
1 ½ cups olive oil
¾ cup vinegar
  Salt and pepper
1 green plantain
1 medium-ripe plantain
1 very ripe plantain
2 tablespoons lemon juice
1 pound *malangas*, peeled and
  cubed

Soak the codfish overnight or for 8 or more hours, changing the water two or three times. If fillets are not available, bone the fish the following day. Parboil the cod for 30 minutes or until it turns flaky. Remove, drain, and place on a platter.

Cut the cod into several long slices and smother with sliced onions, tomatoes, roasted peppers, and whole garlic cloves. Cover with 1 cup of oil and the vinegar. Season with salt and pepper. Leave to marinate at room temperature for at least 6 hours.

Wash the plantains and slice into 3-inch chunks. Place in a bowl of water with lemon juice. Boil the *malangas* in salted water. Add the plantain slices 10 minutes after the water rolls. Boil for 20 minutes or until all the vegetables are tender. Remove from the water and drain. Peel the plantains and toss with the *malangas* in the remaining ½ cup of oil. Arrange on a platter around a centerpiece of fish. Serve while the vegetables are warm.

Serves 6

Salads

BANANA CRAB CUPS

2 medium bananas
1 tablespoon lemon juice
1 cup cooked crab meat
½ cup chopped celery
¼ cup dill pickle, finely chopped
3 tablespoons mayonnaise
1 tablespoon Dijon or French-
  style mustard
4 large, ripe avocados
1 tablespoon fresh parsley,
  chopped
Lettuce

Peel the bananas, slice into small cubes, and sprinkle with lemon juice. Combine in a mixing bowl with the crab meat, celery, and pickle. Blend the mayonnaise and mustard and mix in. Peel the avocados and cut in half. Remove the seeds and cut off a thin slice from the bottom of each avocado so it will stand easily. Spoon the banana-crab mixture into each avocado and garnish with parsley. Set the avocados on lettuce leaves. Allow 2 per person.

Serves 4

BANANA TURKEY

8 large bananas
  Juice of 3 lemons
½ cup cooked turkey meat
2 stalks celery, chopped
1 cup mayonnaise
1 teaspoon paprika
  Salt and pepper
1 tablespoon parsley, chopped

This recipe is perfect for using Thanksgiving dinner leftovers.

Make a lengthwise gash in each banana peel large enough to extract the banana without tearing the peel. Remove the bananas and dice finely, sprinkling with lemon juice. Also sprinkle the insides of the peels. Dice the turkey meat and combine with the celery, mayonnaise, paprika, and salt and pepper to taste. Mix with the diced bananas. Place the mixture by spoonfuls in the empty banana peels until full. Garnish with parsley.

Serves 4

BANANA WATERCRESS SALAD

Bunch of watercress
6 bananas
¼ cup orange juice
2 tablespoons lemon juice
¼ cup sour cream
1 tablespoon milk
2 tablespoons mayonnaise
2 teaspoons sugar
1 teaspoon paprika
Salt and pepper
2 radishes, sliced

Wash the watercress, removing any large stems, and place in a salad bowl. Peel the bananas, slice lengthwise, and lay crisscross on the watercress. Sprinkle the fruit juices over the banana slices. In a mixing bowl, combine the sour cream, milk, mayonnaise, sugar, paprika, and salt and pepper to taste. Mix well and pour over the bananas. Make a ring of radish slices in the center of the bowl. Chill.

Serves 6

BANANA FRUIT SALAD

4 navel oranges
4 ripe bananas
2 tablespoons lime juice
1½ cups chopped pineapple or canned unsweetened pineapple chunks
4 peeled peaches, diced
1 cup seedless grapes
½ cup slivered almonds
1 cup milk
1 tablespoon cornstarch
2 tablespoons butter
½ cup sugar
1 cup whipping cream

Peel the oranges and divide into sections. Peel the bananas and slice into rounds, sprinkling with lime juice to prevent their turning brown. Combine the bananas, oranges, pineapple, peaches, grapes, and almonds in a bowl, toss, and chill.

In a saucepan, heat the milk to scalding. Mix the cornstarch with 1 tablespoon cold milk and pour into the saucepan. Add the butter and sugar and stir. Cook over medium heat until the sauce thickens somewhat. Chill for 30 minutes and stir in the whipped cream. Pour the sauce over the fruit and serve.

Serves 6

BANANA PINEAPPLE SALAD

2 cups canned unsweetened pineapple chunks
  Pineapple juice
1 tablespoon cornstarch
2 egg yolks
2 tablespoons butter
3 ripe bananas
2 tablespoons lemon juice
1 cup seedless grapes
1 cup walnuts
2 cups shredded cabbage

Pour the liquid from the canned pineapple into a saucepan. Add enough pineapple juice to make 1 cup. Mix the cornstarch with 1 tablespoon juice and add to the saucepan. Cook over medium heat. Add the egg yolks, beaten, and the butter. Stir until the mixture becomes slightly thick. Chill.

Peel the bananas, slice into rounds, and sprinkle with lemon juice to prevent their turning brown. Combine the bananas with the pineapple, grapes, walnuts, and cabbage. Pour the dressing over the salad and toss. Chill for 1 hour before serving.

Serves 6

BANANA APPLE SALAD

4 ripe bananas
6 apples, diced
2 tablespoons lemon juice
½ cup seedless grapes
½ cup chopped celery
¼ cup grated coconut
½ cup slivered almonds
½ cup whipping cream
½ cup mayonnaise
½ teaspoon almond extract
¼ cup superfine sugar
  Lettuce leaves

Peel the bananas and slice into rounds. Combine with the diced apples and toss with lemon juice. Add the grapes, celery, coconut, and almonds. Whip the cream to a medium consistency and mix with the mayonnaise, almond extract, and sugar. Pour the mixture over the apples and bananas, toss, and chill. Serve on lettuce leaves.

Serves 6

BANANA WALDORF SALAD

1 ½ cups mayonnaise
2 tablespoons lemon juice
1 teaspoon paprika
4 apples, peeled and diced
1 cup celery, diced
1 cup walnut pieces
1 cup seedless grapes
4 ripe bananas
1 head iceberg lettuce

All ingredients should be chilled. Mix the mayonnaise with the lemon juice and half the paprika. Add the apple chunks, coating them well. Gradually add the celery, walnuts, and grapes. Peel the bananas, cut into 1-inch-thick slices, and add to the mixture. Spread leaves of iceberg lettuce on individual plates and spoon the salad over. Dust with the remaining paprika.

Serves 8 to 10

MANILA SALAD (PHILIPPINES)

4 ripe bananas
2 tablespoons lemon juice
2 mint leaves
¼ cup mayonnaise
2 tablespoons Dijon or French-style mustard
2 teaspoons salt
1 teaspoon white pepper
1 cup diced celery
1 cup flaked salmon
Lettuce leaves
4 tomatoes, sliced in wedges
1 teaspoon paprika

Peel the bananas and cut into 1-inch-thick rounds. Sprinkle with lemon juice to prevent their turning brown. Chop up the mint leaves and toss with the bananas. Mix the mayonnaise with the mustard, salt, and pepper. Stir in the bananas, celery, and salmon. Spoon into a salad bowl lined with lettuce leaves. Place tomato wedges around the border and dust with paprika.

Serves 6 to 8

Sweets

UNITED FRUIT FREEZE (COSTA RICA)

6 ripe bananas
¼ cup lime juice
2 tablespoons sugar
1 3-inch stick cinnamon
3 cups water
¼ cup orange juice
2 egg whites

Peel the bananas and mash with the lime juice. Place in a saucepan and add the sugar, cinnamon, water, and orange juice. Simmer for 10 minutes. Remove from heat and let cool. Beat the egg whites until they form soft, moist peaks and fold into the mixture. Pour into trays and freeze.

Serves 4

BROILED BANANAS

6 green-tipped bananas
¼ cup melted butter
Juice of 2 limes

Wash the bananas, dry, and place in a broiler. Broil 4 inches from flame for about 10 minutes, turning twice, until the skins are almost black. Remove, peel, and set on a serving plate. Smother with butter and sprinkle with lime juice. Serve with roast pork.

Serves 6

BRAISED BANANAS

8 ripe bananas
¼ cup sweet butter
½ cup water
1 vanilla bean
¼ cup brown sugar
½ cup grated Cheddar cheese

Peel the bananas, cut into 2-inch slices, and brown quickly in butter. Add the water, vanilla, and sugar. Cover and simmer, basting the bananas until the syrup is thick. Before serving, sprinkle grated Cheddar cheese on top.

Serves 8

HAWAIIAN BANANA ROLLS

4 greenish bananas
Juice of 1 lemon
½ cup shredded coconut
½ cup whipped cream
½ teaspoon cinnamon

Peel the bananas, sprinkle with lemon juice, and roll in the shredded coconut. Place in a buttered baking dish and bake in a preheated 350-degree oven for 10 to 15 minutes, or until the coconut turns brown. Serve with whipped cream sprinkled with cinnamon.

Serves 4

BANANA SHERBET

4 large, ripe bananas
4 tablespoons lemon juice
4 cups water
1 cup superfine sugar
2 tablespoons unflavored gelatin

Peel and mash the bananas, adding lemon juice to prevent discoloring. Boil 3 cups of water and pour in the sugar, stirring, and simmer for 30 minutes to form a heavy syrup. Dissolve the gelatin in 1 cup water, add to the syrup, and stir. Strain and cool until not quite firm. Fold in the mashed bananas, pour into metal trays, and freeze at least 4 hours.

Serves 8 to 10

BANANA ICE CREAM I

2 cups milk
2 cups cream
8 egg yolks
1 teaspoon vanilla extract
1 cup superfine sugar
¼ teaspoon salt
2 tablespoons banana liqueur
6 ripe bananas
1 tablespoon lime juice

Mix the milk and cream and bring to the scalding point; then let cool somewhat. Beat the egg yolks with the vanilla, sugar, and salt until the mixture is fluffy. Stir the yolks into the milk and cook in the top of a double boiler until the custard thickens. Stir in the banana liqueur. Cool for 1 hour.

Peel and mash the bananas, adding lime juice to prevent discoloring. Add to the custard, stirring thoroughly. Pour the mixture into 2 1-quart shallow metal trays and freeze at least 4 hours or until firm.

Serves 6

BANANA ICE CREAM II

4 eggs
1 cup sugar
5 cups milk, or 2½ cups milk
　　plus 2½ cups cream
4 ripe bananas
2 tablespoons white rum
½ teaspoon salt

Beat the eggs thoroughly. Add the sugar, and the milk or milk and cream. Cook in a double boiler for about 4 minutes. Peel the bananas and mash with the rum. Add the bananas gradually to the custard. Add the salt, stir, and pour into metal trays. Freeze for about 4 hours.

Serves 6 to 8

BANANA MEN

8 slices pineapple
8 lettuce leaves
8 bananas
2 tablespoons lemon juice
½ cup shredded coconut
2 tablespoons seedless raisins
8 ripe strawberries

This recipe is recommended for children's parties, birthdays, and similar occasions.

Set each pineapple slice on a leaf of lettuce. Peel the bananas, halve crosswise, and sprinkle with lemon juice. Take 8 banana halves and cut each into quarters. Roll all the pieces in coconut. Place the whole banana halves atop each pineapple ring. Using the quartered halves, make legs and arms around each pineapple slice. Use three raisins for each banana half to make a face for the figure. Set a strawberry in the center of each pineapple slice and serve.

Serves 8

BANANA WHIP

1½ cups mashed bananas (about
　　2 bananas)
　　Juice of 1 lemon
⅓ cup light rum
¼ cup sugar
1 teaspoon lemon rind
3 egg whites
¼ teaspoon nutmeg
　　Whipped cream (optional)

Combine the mashed bananas with the lemon juice, rum, sugar, and lemon rind. Beat the egg whites until they form stiff peaks. Fold whites carefully into the banana mixture. Sprinkle with nutmeg. Chill 4 to 5 hours. Serve with whipped cream if desired.

Serves 4

ORANGE BANANA CREAM

6 bananas
3 cups orange juice
1/4 cup superfine sugar
1/2 teaspoon cinnamon
1/2 cup whipping cream

Peel the bananas and mash, gradually adding the orange juice, sugar, and cinnamon. When the mixture is smooth, beat the cream until stiff and fold in. Chill 2 to 3 hours. Serve over hot gingerbread.

Serves 4 to 6

HEAVENLY BANANAS (GUADELOUPE)

1 8-ounce package cream cheese
1/4 cup chopped walnuts
1/3 cup superfine sugar
1 teaspoon cinnamon
1/3 cup sweet butter
6 ripe bananas
3 egg whites
1 teaspoon salt

In a mixing bowl, blend together well the cream cheese, walnuts, 4 tablespoons sugar, and 1/2 teaspoon cinnamon.

Melt the butter in a frying pan. Peel the bananas, slice lengthwise, and sauté until golden brown. Lay half the bananas on the bottom of a greased ovenproof dish. Spread half the cream cheese mixture over the bananas. Place the remaining bananas on top and cover with the rest of the cheese mixture.

Beat the egg whites with the remaining sugar and salt until they form soft, moist peaks. Spread over the cheese topping. Place the dish in a preheated 325-degree oven for about 20 minutes or until the meringue is a golden color. Sprinkle with the rest of the cinnamon. Serve hot.

Serves 6

BANANA DAUBE (ST. MARTIN)

4 large bananas
1 tablespoon lime juice
1/4 cup butter
1 cup red wine
1/2 teaspoon allspice
1/4 cup sugar

Peel and slice each banana crosswise into four parts. Sprinkle with lime juice. In a frying pan, sauté the bananas in butter until golden brown on both sides. Combine the red wine, allspice, and sugar in a separate saucepan. Heat until the sugar dissolves and pour over the bananas. Cook the bananas in the liquid over very low heat for about 15 minutes. Serve warm with vanilla ice cream on the side.

Serves 4

BABY BANANA FRITTERS

12 baby bananas or 6 bananas
   halved crosswise
3 tablespoons sugar
¾ cup gold rum
1 vanilla bean

Batter:
3 cups flour
1½ teaspoons baking powder
2 tablespoons sugar
1 teaspoon salt
1 teaspoon allspice
3 egg yolks, beaten
1 cup milk
2 tablespoons melted butter
   Oil for frying

For best results, use the tiny bananas, 3 to 4 inches long, sold in fine food stores. If they are not available, slice medium-sized bananas in half.

Peel the bananas and steep in a mixture of 3 tablespoons sugar, half the rum, and the vanilla bean for 30 minutes, turning once.

Prepare the batter by sifting together the flour, baking powder, 2 tablespoons sugar, salt, and allspice. Add the remaining rum, egg yolks, milk, and butter gradually, stirring to a medium consistency. Dip the bananas in the batter and deep-fry over medium-high heat until golden brown. Remove, drain, and sprinkle with confectioners' sugar. Serve warm.

Serves 6

FLAMING BANANAS

6 large bananas
¾ cup brandy
¾ cup sugar
   Juice of 1 lemon
1 teaspoon vanilla extract
2 teaspoons cinnamon
⅓ cup sweet butter

Peel the bananas and halve lengthwise. Place on a buttered ovenproof serving dish. Combine half the brandy with the sugar, lemon juice, vanilla, and cinnamon and pour over the bananas. Melt the butter and drizzle over the bananas. Bake at 375 degrees for 20 minutes, basting frequently. Before serving, warm the remaining brandy, pour over the bananas, and flame.

Serves 6

BANANA PARFAIT CHIFFON

½ teaspoon salt
½ cup sugar
1 envelope unflavored gelatin
2 eggs, separated
½ cup milk
½ cup cream
3 medium bananas, mashed
1 teaspoon vanilla extract
1 tablespoon rum
   Whipped cream (optional)

In a saucepan, combine the salt, sugar, and gelatin. Add the egg yolks, milk, and cream while beating briskly with a wire whisk. Heat over a low flame, stirring constantly, until the mixture starts to boil. Remove from heat, add the mashed bananas, vanilla, and rum, and stir. Chill the mixture until it mounds on a spoon.

Beat the egg whites until they form soft, moist peaks. Fold into the banana mixture. Pour into individual cups and chill for at least 1 hour until the gelatin sets. Garnish with whipped cream, if desired.

Serves 8

BANANA TAPIOCA

2 tablespoons butter
2 ripe bananas, mashed
¼ cup quick-cooking tapioca
1¾ cups milk
1 egg, separated
¼ teaspoon salt
¼ cup sugar
1 teaspoon nutmeg

Melt the butter in a saucepan and add the mashed bananas, stirring constantly for 3 minutes. Set aside.

In another saucepan, combine the tapioca, milk, egg yolk, salt, and sugar. Cook the tapioca for 6 to 8 minutes over medium heat, stirring constantly. Cool.

Beat the egg white until it forms crests but is not stiff. Fold the banana into the beaten egg white, and this, in turn, into the tapioca mixture. Blend well. Pour into a bowl, sprinkle with nutmeg, and chill.

Serves 6

FLAMBÉED BANANA CRÊPES (BELGIUM)

Batter:
½ cup milk
¼ cup butter
1 tablespoon sugar
1 teaspoon salt
1⅓ cups flour
2 eggs
2 teaspoons oil
½ cup flat beer

Filling:
3 ripe bananas
1 cup whipping cream
2 tablespoons lemon juice
2 tablespoons sugar
¼ cup confectioners' sugar
¼ cup dark rum

In a saucepan, heat together the milk, butter, sugar, and salt until the butter is melted. Sift the flour into a mixing bowl. Make a well in the center and add the eggs and oil. Beat thoroughly with a wire whisk, then add the milk and beer. Let the batter stand for 2 hours before using.

Using a crêpe pan, make the crêpes and stack them on a plate.

Peel and mash the bananas, and mix with the whipping cream, lemon juice, and sugar. Take one crêpe at a time and spread with the banana mixture. Fold. Place the filled crêpes on a warm serving dish until all crêpes have been filled. Sprinkle the crêpes with confectioners' sugar. Warm up the rum, pour over the crêpes, and flame without letting the crêpes brown. Serve immediately.

12 crêpes

BANANA GRIDDLE CAKES (VENEZUELA)

4 ripe bananas
Juice of 1 lemon
½ tablespoon salt
2 tablespoons baking powder
2 tablespoons vegetable shortening
4 eggs, separated
2 cups milk
2 cups flour
½ tablespoon nutmeg
2 tablespoons sugar

Mash the bananas and sprinkle with lemon juice to prevent darkening. Add the salt, baking powder, and shortening to the egg yolks. Beat well. Add the milk and beat again. Let stand 30 minutes. Mix in the bananas, flour, nutmeg, and sugar.

Beat the egg whites until they form soft, moist peaks. Fold them into the batter. Fry by the spoonful on a griddle, turning once. Serve hot with honey or guava jelly.

About 3 dozen cakes

BANANA COOKIES

½ cup sweet butter
¼ pound brown sugar
2 ripe bananas
1 egg, beaten
½ cup seedless raisins
1 ¾ cups flour
½ teaspoon salt
1 teaspoon baking powder
½ teaspoon vanilla extract

Cream the butter and sugar together. Peel and mash the bananas. Add gradually to the butter mixture, alternating with the beaten egg and raisins.

Sift together the flour, salt, and baking powder and fold into the batter. Add the vanilla and mix well. Let stand 30 minutes. Drop by the spoonful onto a greased cookie sheet. Bake in a preheated 350-degree oven for 15 minutes.

About 30 cookies

BANANA OATMEAL COOKIES

2 cups flour
1 teaspoon baking powder
¼ teaspoon salt
⅔ cup butter
½ cup sugar
1 large, ripe banana, mashed
1 egg, beaten
⅓ cup milk
½ teaspoon vanilla
2 cups rolled oats
⅓ cup raisins
¼ cup walnuts

Sift together the flour, baking powder, and salt. Cream the butter and sugar. Add the mashed banana, egg, and milk and mix until creamy. Combine well with the remaining ingredients. Drop by the spoonful onto a greased cookie sheet. Bake in a preheated 350-degree oven for 15 to 20 minutes.

About 60 cookies

BANANA CRISP

8 medium, ripe bananas
2 tablespoons lemon juice
½ cup butter
¾ cup vanilla wafer crumbs
½ teaspoon cinnamon
½ teaspoon vanilla extract
Grated rind of 1 lemon
½ cup brown sugar
½ teaspoon salt

Peel the bananas and slice in half lengthwise. Sprinkle with lemon juice and place on a buttered baking dish. Melt the butter and combine with the vanilla wafer crumbs, cinnamon, vanilla extract, lemon rind, sugar, and salt. Mix into a paste and spread over the bananas. Place the dish in a preheated 350-degree oven. Bake for 30 minutes or until crisp on top. Garnish each serving with a scoop of vanilla ice cream.

Serves 8 to 10

BANANA RICE BREAD

½ cup sweet butter
1 cup milk
2 medium, ripe bananas
1 cup uncooked cream of rice
2 teaspoons baking powder
½ cup sugar
¼ teaspoon salt
1 teaspoon nutmeg
¼ cup currants

Melt the butter in a saucepan and combine with the milk. Peel the bananas, cut into rounds, and blend with the milk in an electric blender. Combine the cream of rice, baking powder, sugar, salt, nutmeg, and currants in a bowl. Gradually pour in the banana milk, beating vigorously until well blended. Pour into a greased loaf pan and bake in a preheated 350-degree oven for 45 minutes.

Serves 6

BANANA PUDDING

6 ripe bananas
  Juice and grated rind of 2
  lemons
4 cups plain bread crumbs
¼ cup seedless raisins
1¾ cups sugar
3 tablespoons melted butter
4 beaten eggs
2 cups milk

Peel the bananas and slice into rounds. Place one-third of the banana slices on the bottom of a buttered baking pan. Sprinkle one-third of the lemon juice over the bananas and cover with 1½ cups bread crumbs and one-third of the raisins. Dust with ¼ cup sugar and one-third of the lemon peel. Baste. Repeat twice, until bananas, bread crumbs, lemons, and raisins are used up.

Mix together the butter, 1 cup sugar, eggs, and milk. Pour over the pudding. Place the baking pan inside a larger pan holding water. Bake in a preheated 350-degree oven for 35 minutes or until a toothpick inserted in the middle of the pudding comes out clean. If needed, add water to the larger baking pan during cooking so that the smaller pan is always surrounded by liquid. Remove from the oven, let cool 10 minutes in the pan, and turn out.

Serves 12

BREAD PUDDING (HAITI)

3 ripe bananas
  Juice and rind of 1 lemon
2 cups stale white bread crumbs
½ cup currants
2½ cups milk
¼ cup dark rum
1 egg, beaten
1 teaspoon vanilla extract
½ cup sugar
1 teaspoon allspice

Peel the bananas and slice lengthwise, coating each side with lemon juice. Soak the bread crumbs and currants in ½ cup milk and half the rum for 30 minutes. Mix in the grated lemon peel. Beat the egg with the rest of the milk and rum and the vanilla, sugar, and allspice.

In a greased baking mold, first place a layer of bread crumbs, then one of bananas, alternating until the ingredients are used up and finishing with a layer of bread crumbs on top. Pour over it the milk and egg mixture. Dot with butter on top and bake in a preheated 325-degree oven for 40 minutes.

Serves 8

KINGSTON BANANA LOAF (JAMAICA)

¼ pound sweet butter
½ cup brown sugar
3 egg yolks
2 cups flour
1 teaspoon ground cloves
1 tablespoon baking powder
Pinch of salt
1 cup mashed banana
1 teaspoon vanilla extract
½ cup currants
½ cup chopped peanuts

Cream the butter and sugar together until fluffy. Add the egg yolks and mix thoroughly. In another bowl, sift together the flour, cloves, baking powder, and salt.

Mix the banana with the vanilla. Add to the egg and butter mixture gradually, alternating with the sifted ingredients. Beat thoroughly until everything is well blended. Add the currants and peanuts to the batter. Mix well and pour into a greased 9-by-5-by-3-inch loaf pan. Bake in a preheated 325-degree oven for 1 hour.

Serves 6

TANTE MARIE'S CAKE (HAITI)

3 large bananas
2 tablespoons lemon juice
¾ cup dark rum
2 pounds sweet potatoes, cooked
3 egg yolks
1½ cups brown sugar
1 cup cream
¼ cup butter, melted
¼ cup raisins
1 teaspoon grated lemon peel
½ teaspoon allspice
Whipped cream

Peel and mash the bananas. Mix with the lemon juice and ¼ cup rum and chill. Mash the sweet potatoes and blend with the bananas and all the other ingredients. Pour into a buttered 9-by-5-inch baking pan and bake in a preheated 325-degree oven for 2 hours.

Remove the cake from the oven, cool, and turn onto a serving platter. Pour the remaining rum on the cake and let it stand at room temperature for about 2 hours. Cut into slices, sprinkling with more rum, if desired. Garnish with whipped cream.

Serves 6 to 8

BANANA MOKA CAKE

1½ cups brown sugar
⅔ cup unsalted butter, melted and cooled
2 eggs
2 medium, ripe bananas
4 squares bitter chocolate
½ teaspoon salt
1 teaspoon baking soda
2 cups flour
1 cup milk
1 teaspoon vanilla extract
¼ cup walnuts

Cream the sugar and butter, adding 1 egg at a time. Mix well. Peel the bananas, mash, and beat in. Melt the chocolate and also beat in. Sift together the salt, soda, and flour. Add gradually to the chocolate mixture, alternating with the milk, vanilla extract, and walnuts. Blend thoroughly. Pour into a buttered cake pan. Bake in a preheated 350-degree oven for 30 minutes. Remove, cool on a rack, and decorate with Banana Icing (see page 160).

Serves 8

BANANA SCONES

2 cups flour
1½ teaspoons baking powder
½ teaspoon salt
⅓ cup vegetable shortening
1 egg
3 tablespoons brown sugar
½ cup milk
½ cup mashed bananas (about 1 banana)
½ cup raisins

Sift together the flour, baking powder, and salt into a large mixing bowl. Add the shortening, rubbing all the ingredients together with your fingers until the mixture becomes like fine crumbs. To make the scones lighter, toss the flour while rubbing.

Beat the egg with the sugar and pour into the mixture. Mix well. Gradually add the milk, bananas, and raisins. Drop the batter by spoonfuls into large muffin tins or molds, filling them halfway. Bake in a preheated 450-degree oven for about 15 minutes or until browned on top.

About 20 scones

INDIAN BANANA CRULLERS

2¼ cups flour
½ tablespoon cinnamon
2 eggs
¼ cup milk
¼ cup mashed bananas (about
  half a banana)
½ tablespoon baking powder
1 cup superfine sugar
  Oil for frying

Sift together the flour and cinnamon into a bowl. Beat the eggs and add to the flour, mixing in the milk gradually until a stiff dough is formed. Add the mashed bananas, baking powder, and half the sugar. Mix well. Roll out on a floured board to 1-inch thickness. Cut the dough into 3-inch-long slices, twist twice, and dip in the remaining sugar. Deep-fry in hot oil for about 2 minutes or until golden brown. Remove, drain on paper towels, and sprinkle with sugar.

About 20 crullers

PASTELILLOS DE PLÁTANO (CUBA)
TINY BANANA PIES

Dough:
1 cup flour
¼ cup shortening
3 tablespoons ice water
½ teaspoon salt

Filling:
2 ripe bananas
  Juice of 1 lemon
½ cup superfine sugar
2 tablespoons cinnamon

Prepare the dough as in Pastry II (see page 160), using instead the amounts listed here.

Roll out the dough to ¼-inch thickness. Cut into oblong shapes about 6 by 4 inches. Peel the bananas, slice into rounds, and sprinkle with lemon juice. Dip in sugar and place 4 or 5 rounds in the middle of each piece of pastry. Fold over the pastry to make a crescent shape. Press the edges together with the tines of a fork. Fry the pies in hot oil for about 3 minutes, turning twice, until golden brown. Remove, drain, and sprinkle with the remaining sugar and cinnamon.

About 10 pies

BANANA CREAM PIE

1 9-inch pie shell, baked (Pastry I or II, pages 159–60)
6 ripe bananas

Filling:
¾ cup superfine sugar
¼ teaspoon salt
5 tablespoons flour
2½ cups milk
3 egg yolks, beaten
2 teaspoons vanilla extract

Topping:
1 cup whipping cream
1 tablespoon sugar
2 ripe bananas

To make the filling, sift together the sugar, salt, and flour into a saucepan. Gradually stir in the milk over low heat until it has a creamy consistency. Simmer 2 or 3 minutes until thickened. Remove 1 or 2 tablespoons of the mixture and beat in with the egg yolks. Add the yolks to the saucepan. Cook over very low heat for 2 or 3 minutes, stirring constantly. Remove from heat and stir in the vanilla. Cool.

Peel the bananas, cut into rounds, and place on the bottom of the pie shell. Cover with the cream filling. Chill 2 to 3 hours. Beat the whipping cream with sugar until it forms peaks. Garnish the top of the pie with whipped cream and two bananas sliced in rounds. Chill.

Serves 8

BANANA TART

1 9-inch pie shell, baked (Pastry I or II, pages 159–60)
4 large bananas
½ cup superfine sugar
Pinch of salt
2 tablespoons butter
Juice of 1 lime
½ teaspoon nutmeg
Whipped cream (optional)

Peel and mash the bananas. Press the pulp through a sieve into a saucepan with the sugar, salt, and butter. Stir and cook over medium heat until the mixture starts to boil. Cool completely, then whip in the lime juice and nutmeg. Pour into the pie shell and chill. Garnish with whipped cream, if desired.

Serves 6

CANDIED PLANTAINS

2 large, medium-ripe plantains
3 tablespoons butter
1 ½ cups brown sugar
1 cup water
⅓ cup fruity white wine
2 cinnamon sticks
6 whole cloves

Peel the plantains and make four diagonal slashes on each side. Melt the butter in a large saucepan and brown the plantains for about 10 minutes, turning frequently. Sprinkle sugar over the plantains and pour in the water and wine. Stir to dissolve the sugar, place the cinnamon sticks and cloves in the pan, and boil for about 10 minutes. Simmer for about 20 minutes, stirring frequently, or until the syrup thickens to the desired consistency. Place on a serving platter and chill. Serve with cream cheese.

Serves 4

CREAMED PLANTAINS

4 medium, ripe plantains
⅓ cup unsalted butter
¼ pound American cheese, cut in matchstick slices
1 cup milk
1 ½ cups cream
¾ cup superfine sugar
1 teaspoon vanilla extract

Peel and wash the plantains. Make a deep lengthwise incision in each plantain, being careful not to split it in half. Melt the butter in a large saucepan and brown the plantains for 4 to 5 minutes, turning them once. Remove the plantains and let cool.

When cool enough to handle, stuff the plantains with cheese slices and replace in the saucepan. Beat together the milk, cream, sugar, and vanilla. Pour over the plantains, bring to a boil, and cover. Simmer for 15 minutes, shaking the pan instead of stirring. Uncover the pan and simmer for 30 minutes, basting the plantains frequently until tender.

Serves 8

SAMOAN COCONUT PLANTAINS

8 small, medium-ripe plantains
¼ cup unsalted butter
1 cup shredded coconut
3 tablespoons superfine sugar
1 ½ cups cream
Whipped cream (optional)

Cook the whole, unpeeled plantains in boiling water for 25 minutes or until tender. Remove, drain, and peel. Halve lengthwise and place in a serving dish.

In a saucepan, melt the butter and brown the coconut for a few minutes. Add the sugar, stir, then add the cream. Cook until the sauce has thickened and pour it over the plantains. Chill. Top with whipped cream, if desired.

Serves 8

Drinks

BANANA CIDER

2 ripe bananas
3 cups sweetened apple cider
½ cup toasted unsalted almonds
¼ teaspoon nutmeg
2 jiggers Calvados (optional)
6 ice cubes

Peel the bananas, slice into rounds, and place in a blender with the rest of the ingredients. Blend at high speed until smooth and frothy.

Serves 6

HAVANA GUABANA

2 large, ripe bananas
1 cup guava juice
2 tablespoons sugar (use only if guava juice is unsweetened)
6 ice cubes

Peel and slice the bananas. Mix with other ingredients in a blender until smooth.

Serves 4

BANANA STRAWBERRY SHAKE

2 large, ripe bananas
8 large, ripe strawberries
2 cups milk
2 tablespoons sugar
4 ice cubes

Peel and slice the bananas. Mix with the other ingredients in a blender at high speed.

Serves 4

TIGER'S MILK

1 banana, sliced
1 cup *kaffir*, or liquid yogurt
1 egg
½ cup orange juice
4 tablespoons brewer's yeast

Combine all the ingredients in a blender until smooth and thick.

Serves 2

BANANA MILKSHAKE

2 ripe bananas
1 egg
4 cups milk
1 teaspoon vanilla extract
¼ cup superfine sugar
½ teaspoon cinnamon or nutmeg
½ teaspoon salt

Peel and slice the bananas. Place in a blender with the other ingredients. Mix at high speed and serve foaming in tall glasses.

Serves 4

BANANA DAIQUIRI

2 small bananas, sliced
8 ounces light rum
3 tablespoons lime juice
2 cups crushed ice
2 ounces brandy
2 ounces banana liqueur (optional)

Place all the ingredients in a blender. Mix at high speed until the drink is foamy. Serve in iced champagne glasses.

Serves 4

Supple-mentary Recipes

BLACK BEAN SOUP (CUBA)

1 pound dried black beans
8 cups water
2 large onions (1 chopped, 1
  whole, peeled)
1 head garlic, unpeeled
2 large tomatoes, peeled
2 bay leaves
1 green pepper
2 pimientos, roasted and sliced
½ cup olive oil
1 teaspoon oregano
½ tablespoon sugar
2 tablespoons vinegar
  Salt and pepper

Pick the beans clean of small leaves and stones and soak overnight in water. Rinse the beans the following day and place in a large casserole. Pour in the water and add the peeled whole onion, head of garlic, 1 tomato, and bay leaves. Bring to a boil, cover, and simmer for 2 to 3 hours or until the beans are tender and the tomato has dissolved.

Roast the green pepper, peel, chop, and place in a frying pan. Add the other tomato, pimientos, olive oil, and chopped onion. Simmer for 5 minutes. Mix in the oregano and 1 tablespoon mashed beans. Combine and simmer 2 to 3 minutes. Pour into the casserole. Season with the sugar, vinegar, and salt and pepper to taste. Continue cooking for about 1 more hour, until the beans begin to disintegrate. Serve with steamed white rice.

Serves 8

SAFFRON RICE

1½ cups water
 2 tablespoons butter
½ teaspoon salt
 1 cup long-grain rice
½ teaspoon powdered saffron

In a pot with a tight-fitting lid, bring the water to a boil with the butter and salt. When the water boils, pour in the rice. Sprinkle saffron on the rice and stir with a *fork*. (Stirring with a spoon will break up the grains and make rice come out mushy.) Bring to a boil again, stir, and cover tightly. Simmer for 15 minutes. Uncover and continue cooking over low heat until the rice has fluffed up.

Note: If you like, you may sauté half a small chopped onion in the butter until the onion is soft before pouring water into the pot.

Serves 6

SALSA RANCHERA

3 tablespoons oil
4 large tomatoes, peeled,
   seeded, and chopped
1 large onion, chopped
2 tablespoons vinegar
6 pickled *jalapeño* peppers, finely
   chopped
1 teaspoon chopped *cilantro*
½ teaspoon sugar
   Salt and pepper

Heat the oil in a frying pan and add the tomatoes, cooking them over medium heat until completely dissolved. Add the onion and cook another 5 minutes over low heat. Remove from heat, season with the remaining ingredients, and mix. Store in the refrigerator.

About 3 cups

PASTRY I

½ cup butter
2 cups sifted flour
½ teaspoon salt
⅔ cup ice water

Separate 2 tablespoons butter and let stand at room temperature. Knead the remaining butter and shape into a flat cake about 3 by 3 inches. The butter should be chilled but malleable.

Sift together the flour and salt and rub in the 2 tablespoons butter until evenly spread throughout. In a mixing bowl, combine the flour and water gradually, mixing with a fork until a dough is formed. Knead for about 15 minutes or until the dough becomes springy. Chill 15 minutes.

Roll out the dough on a floured board into a strip roughly 10 inches long, 8 inches wide, and ¼ to ½ inch thick. The long side should be facing you. Take the butter "cake" and place crosswise in the center of the dough. Fold over both long ends and press firmly. Do the same with the sides. Make sure the butter is firmly enclosed. *If butter leaks out, patch with a piece of dough and rechill for 15 minutes.* It is very important that the butter not come out while preparing this recipe.

Give the pastry a half turn, so short sides face you. Roll out the dough again into a strip roughly 10 by 8 inches. Repeat the folding-over operation, again making sure the butter does not escape. Roll out again. Repeat the rolling and folding two more times, or "turns." Chill in the refrigerator for 45 minutes, remove, and repeat the turns 3 more times. There should be a total of 8 turns.

To Bake: Line the inside of a pie pan with pastry. Chill or freeze for 3 hours. Remove and bake in a preheated 450-degree oven for 10 minutes. Lower the temperature to 350 degrees and bake 15 minutes longer. Remove from the oven, cool, and use.

One 9-inch pie shell

PASTRY II

1¼ cups flour
½ teaspoon salt
7 tablespoons shortening
3 tablespoons ice water

Sift together the flour and salt into a bowl. Mix in half the shortening, cutting it in with two knives until the mix resembles fine crumbs. Add the remaining shortening and cut in again until the mix is broken up into granules shaped like small pebbles. Pour in the water gradually and mix with a fork. Make sure all the flour is wet. Press the dough together with your hands. Knead lightly for 10 seconds.

Turn the dough onto a floured board and roll it out from the center to the edges until the dough is approximately ¼ inch thick. It should also be at least 1 inch wider than the pan to be used.

Place the pastry in the pie pan, making sure there are no bubbles of air between dough and pan. Pinch the edges of the dough on the rim of the pie pan, using the tines of a fork. Pierce the dough three or four times with a fork. Bake in a preheated 425-degree oven for 15 minutes. Remove and let cool.

One 9-inch pie shell

BANANA ICING

2½ cups confectioners' sugar
¼ cup mashed bananas (about half a banana)
2 tablespoons lemon juice
3 tablespoons unsalted butter

Sift the sugar and combine with the banana, lemon juice, and butter. Beat until the icing is smooth and creamy. Spread on tarts and cakes.

2 to 3 cups

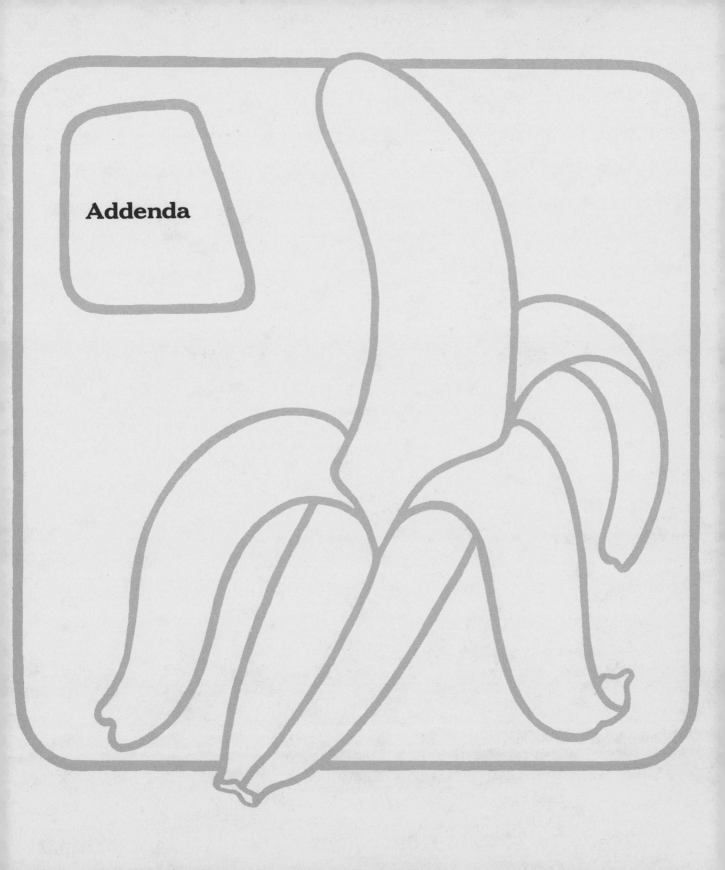

Addenda

BUYING BANANAS (AND PICKING PLANTAINS)

Although you might think all bananas are alike, they are not all of equal quality. Sometimes the pulp has dark spots, is mushy in parts, or lacks the sweetness and texture we favor. How can you learn to avoid these drawbacks without planting a rhizome in the garden? Simple. Just watch out for the peels.

A top banana is plump, unblemished, firm to the touch, and bright to the eye. If your banana doesn't pass muster, it probably suffered some injury during the three to seven thousand miles it traveled between plantation and supermarket.

Although hazards such as beetles and spore diseases can affect the quality of bananas growing on the tree, the major enemy of the banana after it is picked is lack of loving care. Bananas must be handled tenderly because they have sensitive skins. If the peel shows dark spots or is discolored, avoid it. The banana was probably handled roughly and the flesh will also be discolored and mushy. If the peel does not shine or the fruit feels soft, the banana was incorrectly ripened and may have a mealy or tasteless texture. If your bananas pass the quality test, you know you are taking a nice bunch home. The difference between good bananas depends on the color of their skins, which tells you how sweet they are.

In their steamy birthplace in the tropics, bananas are picked, packed, and shipped green. Once they arrive at the local wholesaler, the boxes of banana hands are placed in large ripening rooms. There the bananas are returned to semitropical temperatures and doused with ethylene gas to ripen them uniformly. This process does not alter the flavor of the banana; its use is dictated by the economics of the business, so that wholesalers can prevent the headaches occasioned by some bunches staying green while others spoil when ripened at room temperature. Once the fruit is at the degree of ripeness most craved in your neighborhood, it is forwarded to the retailer.

When displayed at the local market, bananas are usually in the last three stages of ripeness—yellow with green tips, all-yellow, and yellow with brown flecks. In the first stage the banana is the firmest, and in the last, the sweetest. Green-tipped are therefore better for cooking and all-yellow are second best. Bananas with brown or even black flecks should *never* be cooked. They should be eaten raw, and fairly quickly at that, because they have only another two or three days' eating life. Cooking only reduces the already ripe fruit to a syrupy pap.

The fundamental rule to remember when buying bananas is that the riper they are, the sweeter and more nutritious they are. When green, the banana contains a large amount of starches that are indigestible in their raw state. During ripening, the natural enzymes of the banana convert these starches into sugars, which make the banana a soft, palatable food. The enzymes also create certain volatile compounds that give the banana its distinctive aroma.

Green raw bananas can be very astringent because of the tannins present in the pulp. These tannins dissolve when the fruit is mature. The only edible parts of a ripe banana that contain tannins are the filaments that adhere to the flesh. These should be removed before eating, unless you like sweetness with a pucker.

Bananas can be picked green and sold ripe because, unlike deciduous fruits such as oranges, they do not have to be "tree sweet." Bananas ripen better once harvested and in fact do not ripen well at all on the tree. There is no basis to the myth that freshly plucked bananas are sweeter or better tasting. Even before the multinational companies reshaped the banana market, natives of the tropics harvested their bunches fully developed but still green. The fruit would be hung upside down in a shady spot to mature.

Another false story about bananas is that they should never be stored in the refrigerator. The truth is that refrigerating does not damage the flavor or texture of the pulp but only of the skin. What chilling does do is stop the ripening process; you should buy bananas when green-tipped and ripen them at home. Once the banana has reached the degree of ripeness wanted, *then* it may be refrigerated. Bananas will keep several days chilled with no side effect other than the peel turning black. However, they should be returned to room temperature before eating so as to bring out their full flavor. Bananas may also be frozen, in which case the pulp will resemble a banana sherbet.

Not all bananas are yellow or 6 to 8 inches long, as you may find out if you go to specialty food stores. Bananas come in all sizes and in at least one other color, Cuban red. However, banana companies import just one type, the Cavendish variety, because it is resistant to disease and to "blow-downs"—sudden tropical windstorms that can devastate a plantation in a matter of hours.

Two other varieties that can be found in the United States are baby bananas, also called Lady Fingers, and apple bananas, or *manzanos*. Baby bananas, like all bananas in this country, are imported from Latin America, in this instance Venezuela or Honduras. They are 3 to 4 inches long, are thin skinned, and have a creamier texture than the normal banana. The *manzano* is 4 to 6 inches long, is plumper than the Cavendish, and has a mellow flavor reminiscent of a ripe apple. Cuban red is slightly larger than the standard banana and is very rare nowadays, even though according to records it was the first variety of banana to be traded in the United States. These kinds are not distributed widely and usually bear a higher price than the regular Cavendish.

Then there is the plantain. A close relative of the banana, it is considered a vegetable rather than a fruit. It resembles the banana so greatly that the first Europeans who saw it believed it was just an odd banana.

If the discovery of the American potato was the salvation of the Irish, the importation of the Asian plantain was the redemption of Latin Americans. As is the case with potatoes, plantains provide a quick and tasty supply of starch in the diet. In addition, the proceeds of the plantain are staggering by comparison to those of other staple crops. The yield of 1 acre of plantains is 43 times greater than that of potatoes and 133 times greater than that of wheat. And it doesn't take half as much trouble to grow or harvest.

Plantains range considerably in size. There are certain types that look identical to normal bananas when green; others seem gargantuan—over 1½ feet in length. Generally plantains sold in the United States are 10 to 12 inches long and can be differentiated from bananas by the well ridges at the corners, which divide the vegetable into three or four distinct sides. In this country plantains are normally available in Latin American markets.

The major distinction between plantains and bananas is that plantains must *always* be cooked before eating, since the ripening process does not convert all their starch into sugars. Moreover, plantain skins are thick, so there is little danger they can be internally bruised.

Plantains also ripen off the tree; the degree of maturity is judged by their skin tone. There are four general stages—allgreen; medium-ripe, or yellow with brown or black spots; ripe, mostly brown or black with little yellow; and very ripe, or black.

When green and sliced into chips, plantains are light and crunchy, with an unmatched delicate taste. When half ripe or "tone," they are something like a mellow-flavored sweet potato, without the heaviness that vegetable often has. When fully ripened and cooked, they sometimes surpass the raw banana in sweetness and creamy texture.

When completely black and with a slightly wilted skin, the plantain has reached the ultimate point of maturity—*it is not spoiled*. Upon peeling, the pulp appears a golden to reddish color, which indicates a high concentration of sugars. At this stage the plantain should not be cooked overly long or it will dissolve into a pap.

HOW TO PEEL A PLANTAIN

A plantain is harder to peel than a banana but is always easier to pare than a potato. If it is boiled with the skin on, the peel will slide off after cooking. If it is raw, take the following steps to peel it.

With a sharp knife, cut off both ends of the plantain, leaving a little of the pulp core exposed. Make a lengthwise incision down one of the ridges at the corners, being careful not to pierce through to the pulp. Do the same with an adjacent ridge, cutting off an entire side of the plantain. With the edge of the knife, lift off the top part of the divided section and pull down sharply. The entire segment of the peel should come off in one piece. Repeat the operation for the other sides and you will have an entire peeled plantain ready to fry, boil, mash, or cook any way you desire.

When peeling ripe plantains for cooking, you may also make another lengthwise incision in the flesh down to the center. Open up the pulp and extract the core, or heart, which sometimes has small seedlike spots. This practice is optional and was initiated by devout sixteenth-century Spaniards who saw the Sign of the Cross in the spots. Slicing the plantain without coring it meant destroying the cross, a profanation that didn't sit too well with the holy fathers of the Inquisition.

BANANA CALORIE COMPARISONS

QUANTITY	ITEM	CALORIES
1	banana	85
2	chocolate cookies	102
2	Fig Newtons	110
1	macaroon cookie	107
1	oatmeal cookie	86
1 serving ($\frac{1}{16}$ quart)	ice cream	174
1 piece	angel cake ($\frac{1}{10}$ cake)	121
1 piece	pound cake ($\frac{1}{10}$ cake)	149
6–8	cashew nuts	84
8–12	mixed nuts	94
1 tablespoon	peanut butter on one average-size slice white bread	328
1	small milk chocolate bar	302
$\frac{1}{2}$ cup	tuna (in water)	127
$\frac{1}{2}$ cup	cottage cheese (creamed)	120
2$\frac{3}{4}$-inch cube	Cheddar cheese	120

NAMES OF BANANAS

EUROPE
ENGLAND: *Banana*
FRANCE: *Banane; Figue d'Adam; Figue du Paradis*
HOLLAND: *Banaan*
SPAIN: *Banano*
PORTUGAL: *Bananeira*
ITALY: *Banana, Fico di Adamo*
GERMANY: *Banane*

LATIN AMERICA
BRAZIL: *Bananeira*
CENTRAL AND SOUTH AMERICA: *Banana, Guineo, Camburi, Plátano*
GUIANA REGION: *Bacove, Bacobe, Bacooba* (from Portuguese *Pacova*, of African origin, or of Amerindian origin)

WEST INDIES
LEEWARD AND WINDWARD ISLANDS: *Banana, Fig, Silk Fig* (the latter in islands having French elements in their history, i.e., Trinidad, Grenada)
FRENCH ISLANDS: *Banane, Figue*
SPANISH-SPEAKING ISLANDS: *Camburi, Guineo* (except Cuba, where it is *Plátano*)

AFRICA
CANARY ISLANDS: *Plátano*
EGYPT: *Mouz, Maouz, Moz, Mazw, Moaz* (Arabic, from Sanskrit *Moka* or perhaps from southern Arabian town of same name)

EAST AFRICA: three main series of names
1. based on Arabic, i.e., *Moz* (Somali), *Maso* (Swahili), *Mazu* (Wadigo), and perhaps *Ndizi* (Swahili);
2. based on *Toke*, i.e., *Kitoke, Madoge, Matoke*
3. based on *Konde*, i.e., *Ikundu, Ngimda, Mikonde, Maonde*
MADAGASCAR: *Akondre, Fontsy, Ontsy*
WEST AFRICA: *Itoto, Atora, Ditotu, Ikondo, Ekon, Digonde, Liko, Banema, Banama, Banana, Ogede, Ayaba,* and a host of others

ASIA
INDIA: *Pala, Kela, Kadali, Moca, Rambha, Suphala, Kardung, Kait, Kabu, Vazhei, Vazha, Arati, Bale, Vazhapazham, Valapalam*
INDOCHINA: *Chuoi, Kok khone, Touille, Chec*
THAILAND: *Klue*
BURMA: *Nget-pyaw* ("the birds told," which refers to the

story that men first ate the fruit when they saw the birds do so), *Nga*

CHINA: *Tsiu, Chiu, Chiao, Kan-chao, Cha*

PHILIPPINES: *Pisang, Saging, Latundan, Bungulan, Saba*

MALAYA: *Pisang*

INDONESIA: *Gedang, Pisang*

NEW GUINEA: *Ami, Deung, Em, Kwali, Lung, Marafe, Napet, Oere, Opie, Usi, Waniga, Sabulong*

NEW CALEDONIA: *Docile, Bolao, Dopwi, Dupijing, Foexac, Futo, Jui, Meteun, Muji, Newi, Pijino, Pijing, Twine, Wi, Mondgui, Pouin*

SOUTH PACIFIC
FIJI: *Jaina, Vudi*

MICRONESIA: *Atu, Chotda, Dinai, Uch, Ut, Wis*

POLYNESIA: *Huki, Huti, Futi, Fusi, Wuti, Taveli* (western Polynesia); *Meika, Mei'a, Mai'a, Tau Tau* (eastern Polynesia); *Fa'i* (Samoa); *Fe'i* (Tahiti)

HAWAII: *Banana, Lacatan, Maiamaoli, Popoulu*

NORTH AMERICA
UNITED STATES AND CANADA: *Banana*

OTHER MEANINGS OF BANANA

(GLOSSARY)

BANAN: yellow color of a ripe banana, as in "shades of banan and cream."

BANANA: (1) an unskillful boxer; (2) a person born in Japan or of Japanese ancestry who tries to pass for Anglo-Saxon or who slavishly follows Anglo-Saxon customs; (3) U.S. dollar, during 1920s and 1930s; (4) front bumper of automobile in car manufacturer's technical language (1950s); (5) quadroon or one-quarter black woman with high yellow skin tone.

BANANA BACKBONE: a weak, indecisive character.

BANANA BALL: a type of slice in golf.

BANANA BELT: a warm and moist area, sheltered from wind and inclement weather, in an otherwise intemperate climate.

BANANA BIRD: South American and West Indian bird (*Icterus leucopteryx*) that feeds mainly on bananas and other tropical fruits.

BANANA BOAT: (*adj.*) slow, tardy, drifting; (*n.*) (1) an antiquated vessel of little speed with no set arrival time or destination; (2) an invasion barge, an aircraft carrier (military slang of World War II).

BANANA BUNCHING: numbers of persons hanging from the same strap or pole in a bus.

BANANA CHAIR: a tall, oblong cane and/or rattan armchair made in Asia.

BANANA CLIP: the bullet clip of a U.S. Army carbine; term originated during Vietnam conflict (1964–1975).

BANANA DOLLAR: currency in use in Singapore during occupation by Japanese forces during the 1940s; currency of little or no value.

BANANA FLY: a tropical pest (*Drosophila ampelophila*).

BANANA HEAD: fool, dunce, stupid person.

BANANA HITTING: having a series of hits in a row (baseball).

BANANA MULE: a small and wiry mule used on plantations in the South.

BANANA NOSE: a large or long hooked nose; appellation given originally to Eddie Arcaro, a famous jockey.

BANANA OIL: (1) a liquid ester, amyl acetate, having the odor of ripe bananas, used principally in making fruit essences; also employed as a solvent and as a vehicle for applying bronze pigments; (2) nonsense, foolishness, lies.

BANANA PASS: a short forward pass (U.S. football).

BANANA PEEL: a spurious psychotropic drug, popular during the 1960s.

BANANA PEEL, FOOT ON: clumsy, awkward, ungraceful, cumbersome; usage derives from the phrase "a foot in the water bucket and the other on the banana peel" (baseball).

BANANA POLE: a bent pole used for vaulting.

BANANA QUIT: See *banana bird*.

BANANA RACE: in New England states, a fixed horse race (rare).

BANANA REPUBLIC: a politically unstable, corrupt, backward nation; a Spanish-speaking nation; a Central American or Caribbean country; any country where bananas are the main product.

BANANAS: nonsense, tomfoolery; corruption of the Italian surname Bonanno.

BANANA SEAT: a low-slung, oblong motorcycle or bicycle riding seat.

BANANA SHOT: a quick, sliding pass (soccer).

BANANA SLALOM: fast, slippery turns on a downhill race (skiing).

BANANA SPLIT: popular concoction of scoops of ice cream atop a split banana, adorned with chocolate bits, whipped cream, and a cherry.

BANANA STICK: a baseball bat of inferior wood that splinters easily.

BANANA TACTICS: encouraging the opposite party in a dispute to continue negotiations without the first party making a concession or commitment; employed during the Cold War period between the United States and the U.S.S.R. (1948–1972); *syn., salami tactics.*

BANANA WAGON: a yellow taxicab; formerly, a station or beach wagon painted yellow that was used by resorts to meet guests at train stations.

FLYING BANANA: a type of helicopter employed to transport combat troops during the Vietnam conflict (1964–1975); *syn., flying cigar.*

GO, GOES, GOING BANANAS: expression signifying a slipping grasp on reality, insanity; extreme fondness for a particular concept or object, i.e., "I go bananas over dill pickles." Origin unknown; allegedly stems from slang of American soldiers serving in the Philippines during the 1898 Spanish-American War.

SECOND BANANA: supporting comic actor or actress.

THIRD BANANA: stage comic who takes the fall (slapstick).

TOP BANANA: protagonist; most important character in a work or an enterprise; main actor.

RECIPE INDEX